SBAC Math Practice Grade 7

Complete Content Review Plus 2 Full-length SBAC Math Tests

Elise Baniam - Michael Smith

SBAC Math Practice Grade 7
Published in the United State of America By
The Math Notion
Email: info@Mathnotion.com
Web: WWW.MathNotion.com

Copyright © 2020 by the Math Notion. All rights reserved. No part of this publication may be reproduced, stored in a retrieval system, or transmitted in any form or by any means, electronic, mechanical, photocopying, recording, scanning, or otherwise, except as permitted under Section 107 or 108 of the 1976 United States Copyright Ac, without permission of the author.
All inquiries should be addressed to the Math Notion.

ISBN: 978-1-63620-032-3

About the Author

Elise Baniam has been a math instructor for over a decade now. She graduated in Mathematics. Since 2006, Elise has devoted his time to both teaching and developing exceptional math learning materials. As a Math instructor and test prep expert, Elise has worked with thousands of students. She has used the feedback of her students to develop a unique study program that can be used by students to drastically improve their math score fast and effectively.

– SAT Math Workbook
– ACT Math Workbook
– ISEE Math Workbooks
– SSAT Math Workbooks
–many Math Education Workbooks
– and some Mathematics books …

As an experienced Math teacher, Mrs. Baniam employs a variety of formats to help students achieve their goals: she teaches students in large groups, and she provides training materials and textbooks through her website and through Amazon.

You can contact Elise via email at:
Elise@Mathnotion.com

SBAC Math Practice Grade 7

Get the Targeted Practice You Need to Excel on the Math Section of the SBAC Test Grade 7!

SBAC Math Practice Grade 7 is **an excellent investment in your future** and the best solution for students who want to maximize their score and minimize study time. Practice is an essential part of preparing for a test and improving a test taker's chance of success. The best way to practice taking a test is by going through lots of SBAC math questions.

High-quality mathematics instruction ensures that students become problem solvers. We believe all students can develop deep conceptual understanding and procedural fluency in mathematics. In doing so, through this math workbook we help our students grapple with real problems, think mathematically, and create solutions.

SBAC Math Practice Book allows you to:

- Reinforce your strengths and improve your weaknesses
- Practice **2500+ realistic** SBAC math practice questions
- Exercise math problems in a variety of formats that provide intensive practice
- Review and study **Two Full-length SBAC Practice Tests** with detailed explanations

...and much more!

This Comprehensive SBAC Math Practice Book is carefully designed to provide only that **clear and concise information** you need.

WWW.MathNotion.com

… So Much More Online!

✓ FREE Math Lessons

✓ More Math Learning Books!

✓ Mathematics Worksheets

✓ Online Math Tutors

For a PDF Version of This Book

Please Visit WWW.MathNotion.com

Contents

Chapter 1: Whole Numbers 11
   Add and Subtract Integers 12
   Multiplication and Division 13
   Absolute Value 14
   Ordering Integers and Numbers 15
   Order of Operations 16
   Factoring 17
   Great Common Factor (GCF) 18
   Least Common Multiple (LCM) 19
   Divisibility Rule 20
   Answer key Chapter 1 21

Chapter 2: Fractions 25
   Adding Fractions – Like Denominator 26
   Adding Fractions – Unlike Denominator 27
   Subtracting Fractions – Like Denominator 28
   Subtracting Fractions – Unlike Denominator 29
   Converting Mix Numbers 30
   Converting improper Fractions 31
   Addition Mix Numbers 32
   Subtracting Mix Numbers 33
   Simplify Fractions 34
   Multiplying Fractions 35
   Multiplying Mixed Number 36
   Dividing Fractions 37
   Dividing Mixed Number 38
   Comparing Fractions 39
   Answer key Chapter 2 40

Chapter 3: Decimal 45
   Round Decimals 46
   Decimals Addition 47
   Decimals Subtraction 48

Decimals Multiplication 49

Decimal Division...50

Comparing Decimals 51

Convert Fraction to Decimal 52

Convert Decimal to Percent 53

Convert Fraction to Percent 54

Answer key Chapter 3..55

Chapter 4: Exponent and Radicals...58

Positive Exponents 59

Negative Exponents 60

Add and subtract Exponents 61

Exponent multiplication 62

Exponent division 63

Scientific Notation 64

Square Roots 65

Simplify Square Roots 66

Answer key Chapter 4..67

Chapter 5: Ratio, Proportion and Percent... 70

Proportions..71

Reduce Ratio 72

Percent 73

Discount, Tax and Tip 74

Percent of Change...75

Simple Interest 76

Answer key Chapter 5..77

Chapter 6: Measurement 79

Reference Measurement..80

Metric Length Measurement 81

Customary Length Measurement 81

Metric Capacity Measurement 82

Customary Capacity Measurement 82

Metric Weight and Mass Measurement 83

Customary Weight and Mass Measurement 83

Unit of Measurements 84

Temperature 85
Time 86
Answers of Worksheets – Chapter 6 87

Chapter 7: Linear Functions 89
Relation and Functions 90
Slope form 91
Slope and Y-Intercept 91
Slope and One Point 92
Slope of Two Points 93
Equation of Parallel and Perpendicular lines 94
Quadratic Equations - Quadratic Formula 95
Answer key Chapter 7 96

Chapter 8: Equations and Inequality 98
Distributive and Simplifying Expressions 99
Factoring Expressions 100
Evaluate One Variable Expressions 101
Evaluate Two Variable Expressions 102
Graphing Linear Equation 103
One Step Equations 104
Two Steps Equations 105
Multi Steps Equations 106
Graphing Linear Inequalities 107
One Step Inequality 108
Two Steps Inequality 109
Multi Steps Inequality 110
Finding Distance of Two Points 111
Answer key Chapter 8 112

Chapter 9: Transformations 117
Translations 118
Reflections 119
Rotations 121
Dilations 123
Coordinates of Vertices 124
Answers of Worksheets – Chapter 9 125

Chapter 10: Geometry 129

Area and Perimeter of Square 130
Area and Perimeter of Rectangle 131
Area and Perimeter of Triangle 132
Area and Perimeter of Trapezoid 133
Area and Perimeter of Parallelogram 134
Circumference and Area of Circle 135
Perimeter of Polygon 136
Volume of Cubes 137
Volume of Rectangle Prism 138
Volume of Cylinder 139
Volume of Spheres 140
Volume of Pyramid and Cone 141
Surface Area Cubes 142
Surface Area Rectangle Prism 143
Surface Area Cylinder 144
Answer key Chapter 10 145

Chapter 11: Statistics and probability 147

Mean, Median, Mode, and Range of the Given Data 148
Box and Whisker Plot 149
Bar Graph 150
Dot plots 151
Scatter Plots 152
Stem–And–Leaf Plot 153
Pie Graph 154
Probability 155
Answer key Chapter 11 156

SBAC Test Review 159

SBAC Practice Test 1 163
SBAC Practice Test 2 177

Answers and Explanations 191

Answer Key 193
SBAC Practice Test 1 195
SBAC Practice Test 2 201

Chapter 1:
Whole Numbers

Add and Subtract Integers

Find the sum or difference.

1) $(+168) + (+76) =$

2) $(+65) + (-32) =$

3) $217 - 69 =$

4) $(-203) + 179 =$

5) $(-45) + 501 =$

6) $182 + (-265) =$

7) $(-9) + 20 =$

8) $360 - 200 =$

9) $(-10) - (-38) =$

10) $(-67) + (-96) =$

11) $(-143) - 234 =$

12) $1250 - (-346) =$

13) $3 + (-12) + (-20) + (-17) =$

14) $(-28) + (-19) + 31 + 16 =$

15) $(-7) - 11 + 27 - 19 =$

16) $6 + (-20) + (-35 - 24) =$

17) $(+24) + (+32) + (-47) =$

18) $(-35) + (-26) =$

19) $-12 - 17 - 16 - 23 =$

20) $7 + (-21) =$

21) $107 - 80 - 73 - (-38) =$

22) $(20) - (-8) =$

23) $(3) - (5) - (-14) =$

24) $(20) - (6) - (-20) =$

Multiplication and Division

Calculate.

1) $340 \times 8 =$

2) $180 \times 30 =$

3) $(-3) \times 7 \times (-4) =$

4) $-3 \times (-6) \times (-6) =$

5) $12 \times (-12) =$

6) $30 \times (-6) =$

7) $6 \times (-1) \times 5 =$

8) $(-600) \times (-50) =$

9) $(-10) \times (-10) \times 2 =$

10) $165 \times 5 =$

11) $160 \times 80 =$

12) $312 \div 12 =$

13) $(-2,475) \div 3 =$

14) $(-32) \div (-8) =$

15) $384 \div (-24) =$

16) $4,500 \div 36 =$

17) $(-84) \div 2 =$

18) $9,588 \div 6 =$

19) $900 \div (-25) =$

20) $1,680 \div 2 =$

21) $(-81) \div 3 =$

22) $(-1,000) \div (-10) =$

23) $0 \div 250 =$

24) $(-680) \div 4 =$

25) $7,704 \div 856 =$

26) $(-3,150) \div 5 =$

27) $7,268 \div 2 =$

28) $(-48) \div (-4)$

Absolute Value

Simplify each equation below.

1) $|-30| =$

2) $-10 + |-30| + 28 =$

3) $|-48| - |-20| + 12 =$

4) $|-9 + 5 - 3| + |3 + 3| =$

5) $2|2 - 14| + 10 =$

6) $|-6| + |-20| =$

7) $|-36 + 20| + 10 - 9 =$

8) $|-10| - |-23| - 5 =$

9) $|-20| - |-10| + 3 =$

10) $|20| - 28 + |-10| =$

11) $\frac{4|3-6|}{2} =$

12) $|-20 + 9| =$

13) $|-20| \times |5| + 5 =$

14) $|-6| + |-36| + 9 - 3 =$

15) $|-20| + |-20| - 40 =$

16) $13 + |-34 + 15| + |-10| =$

17) $28 - |-63| + 10 =$

18) $\frac{|120|}{|4|} + 6 =$

19) $|-9 + 12| + |32 - 15| + 6 =$

20) $|-20 + 15| + |-5| + 3 =$

21) $\frac{|-32|}{8} \times |-6| =$

22) $\frac{4|4 \times 6|}{2} \times \frac{|-16|}{4} =$

23) $\frac{|2 \times 6|}{12} \times 6 =$

24) $|-10 + 2| \times \frac{|-3 \times 5|}{3} =$

25) $|-100 + 8| - 5 + 5 =$

26) $|-50 + 40| - 10 =$

Ordering Integers and Numbers

Order each set of integers from least to greatest.

1) $7, -8, -5, -2, 3$ ___, ___, ___, ___, ___, ___

2) $-3, -16, 4, 10, 9$ ___, ___, ___, ___, ___, ___

3) $18, -18, -19, 25, -20$ ___, ___, ___, ___, ___, ___

4) $-9, -35, 15, -7, 42$ ___, ___, ___, ___, ___, ___

5) $47, -52, 28, -55, 34$ ___, ___, ___, ___, ___, ___

6) $88, 36, -29, 67, -44$ ___, ___, ___, ___, ___, ___

Order each set of integers from greatest to least.

7) $12, 18, -10, -12, -4$ ___, ___, ___, ___, ___, ___

8) $29, 36, -14 - 26, 69$ ___, ___, ___, ___, ___, ___

9) $75, -26, -18, 47, -7$ ___, ___, ___, ___, ___, ___

10) $58, 72, -16, -12, 94$ ___, ___, ___, ___, ___, ___

11) $-7, 99, -15, -48, 64$ ___, ___, ___, ___, ___, ___

12) $-80, -45, -40, 18, 29$ ___, ___, ___, ___, ___, ___

Order of Operations

Evaluate each expression.

1) $5 + (4 \times 3) =$

2) $12 - (3 \times 5) =$

3) $(16 \times 3) + 10 =$

4) $(15 - 5) - (6 \times 3) =$

5) $22 + (16 \div 2) =$

6) $(16 \times 5) \div 5 =$

7) $(84 \div 4) \times (-2) =$

8) $(9 \times 5) + (35 - 12) =$

9) $60 + (2 \times 2) + 8 =$

10) $(30 \times 5) \div (2 + 1) =$

11) $(-8) + (10 \times 4) + 13 =$

12) $(7 \times 6) - (32 \div 4) =$

13) $(9 \times 8 \div 3) - (10 + 11) =$

14) $(12 + 8 - 15) \times 6 - 3 =$

15) $(30 - 12 + 40) \times (95 \div 5) =$

16) $22 + (20 - (32 \div 2)) =$

17) $(6 + 9 - 5 - 8) + (18 \div 2) =$

18) $(85 - 10) + (10 - 15 + 9) =$

19) $(10 \times 2) + (12 \times 5) - 12 =$

20) $12 + 8 - (32 \times 4) + 30 =$

Factoring

Factor, write prime if prime.

1) 12

2) 26

3) 32

4) 48

5) 60

6) 64

7) 35

8) 30

9) 56

10) 75

11) 25

12) 18

13) 49

14) 15

15) 42

16) 124

17) 56

18) 40

19) 75

20) 20

21) 96

22) 27

23) 72

24) 50

25) 24

26) 88

27) 68

28) 124

Great Common Factor (GCF)

Find the GCF of the numbers.

1) 8, 12

2) 48, 32

3) 42, 18

4) 10, 15

5) 18, 24

6) 16, 12

7) 80, 45

8) 100, 75

9) 64, 8

10) 36, 72

11) 93, 62

12) 15, 90

13) 60, 30

14) 36, 28

15) 18, 45

16) 35, 42

17) 12, 20

18) 90, 120, 20

19) 49, 144

20) 16, 28

21) 14, 8, 21

22) 4, 16, 20

23) 14, 49, 7

24) 21, 12

Least Common Multiple (LCM)

Find the LCM of each.

1) 6, 9

2) 30, 24

3) 8, 4, 6

4) 15, 12

5) 30, 5, 40

6) 45, 15

7) 15, 10, 8

8) 3, 4

9) 10, 20, 25

10) 64, 44

11) 24, 36

12) 108, 64

13) 20, 10, 40

14) 12, 20

15) 45, 9, 3

16) 27, 63

17) 42, 12

18) 20, 45

19) 25, 15

20) 14, 32

21) 16, 18

22) 9, 17

23) 32, 18

24) 16, 12

Divisibility Rule

Apply the divisibility rules to find the factors of each number.

1) 12   2, 3, 4, 5, 6, 9, 10   13) 18   2, 3, 4, 5, 6, 9, 10

2) 326   2, 3, 4, 5, 6, 9, 10   14) 405   2, 3, 4, 5, 6, 9, 10

3) 748   2, 3, 4, 5, 6, 9, 10   15) 945   2, 3, 4, 5, 6, 9, 10

4) 81   2, 3, 4, 5, 6, 9, 10   16) 186   2, 3, 4, 5, 6, 9, 10

5) 891   2, 3, 4, 5, 6, 9, 10   17) 640   2, 3, 4, 5, 6, 9, 10

6) 345   2, 3, 4, 5, 6, 9, 10   18) 150   2, 3, 4, 5, 6, 9, 10

7) 75   2, 3, 4, 5, 6, 9, 10   19) 350   2, 3, 4, 5, 6, 9, 10

8) 450   2, 3, 4, 5, 6, 9, 10   20) 4,520   2, 3, 4, 5, 6, 9, 10

9) 1,325   2, 3, 4, 5, 6, 9, 10   21) 990   2, 3, 4, 5, 6, 9, 10

10) 78   2, 3, 4, 5, 6, 9, 10   22) 368   2, 3, 4, 5, 6, 9, 10

11) 772   2, 3, 4, 5, 6, 9, 10   23) 208   2, 3, 4, 5, 6, 9, 10

12) 162   2, 3, 4, 5, 6, 9, 10   24) 500   2, 3, 4, 5, 6, 9, 10

Answer key Chapter 1

Add and Subtract Integers

1) 244
2) 33
3) 148
4) −24
5) 456
6) −83
7) 11
8) 160
9) 28
10) −163
11) 377
12) 1,596
13) −46
14) 0
15) −10
16) −73
17) 9
18) −61
19) −68
20) −14
21) −8
22) 28
23) 12
24) 34

Multiplication and Division

1) 2,720
2) 5,400
3) 84
4) −108
5) −144
6) −180
7) −30
8) 30,000
9) 200
10) 825
11) 12,800
12) 26
13) −825
14) 4
15) −16
16) 125
17) −42
18) 1,598
19) −36
20) 840
21) −27
22) 100
23) 0
24) −170
25) 9
26) −630
27) 3,634
28) 12

Absolute Value

1) 30
2) 48
3) 40
4) 13
5) 34
6) 26
7) 17
8) −18
9) 13
10) 2
11) 6
12) 11
13) 105
14) 48
15) 0
16) 42
17) −25
18) 36
19) 26
20) 13
21) 24
22) 192
23) 6
24) 40

25) 92   26) 0

Ordering Integers and Numbers

1) −8, −5, −2, 3, 7
2) −16, −3, 4, 9, 10
3) −20, −19, −18, 18, 25
4) −35, −9, −7, 15, 42
5) −55, −52, 28, 34, 47
6) −44, −29, 36, 67, 88
7) 18, 12, −4, −10, −12
8) 69, 36, 29, −14, −26
9) 75, 47, −7, −18, −26
10) 94, 72, 58, −12, −16
11) 99, 64, −7, −15, −48
12) 29, 18, −40, −45, −80

Order of Operations

1) 17
2) −3
3) 58
4) −8
5) 30
6) 16
7) −42
8) 68
9) 72
10) 50
11) 45
12) 34
13) 3
14) 27
15) 1,102
16) 26
17) 11
18) 79
19) 68
20) −78

Factoring

1) 1,2,3,4,6,12
2) 1,2,13,26
3) 1,2,4,8,16,32
4) 1,2,3,4,6,8,12,16,24,48
5) 1,2,3,4,5,6,10,12,15,20,30,60
6) 1,2,4,8,16,32,64
7) 1,5,7,35
8) 1,2,3,5,6,10,15,30
9) 1,2,4,7,8,14,28,56
10) 1,3,5,15,25,75
11) 1,5,25
12) 1,2,3,6,9,18
13) 1,7,49
14) 1,3,5,15
15) 1,2,3,6,7,14,21,42
16) 1,2,4,31,62,124
17) 1,2,4,7,8,14,28,56
18) 1,2,4,5,8,10,20,40
19) 1,3,5,15,25,75
20) 1,2,4,5,10,20
21) 1,2,3,4,6,8,12,24,32,48,96
22) 1,3,9,27
23) 1,2,3,4,6,8,9,12,18,24,36,72
24) 1,2,5,10,25,50
25) 1,2,3,4,6,8,12,24
26) 1,2,4,8,11,22,44,88
27) 1,2,4,17,34,68
28) 1,2,4,31,62,124

Great Common Factor (GCF)

1) 4
2) 16
3) 6
4) 5
5) 6
6) 4

7) 5	13) 30	19) 1
8) 25	14) 4	20) 4
9) 8	15) 9	21) 1
10) 36	16) 7	22) 4
11) 31	17) 4	23) 7
12) 15	18) 10	24) 3

Least Common Multiple (LCM)

1) 18	9) 100	17) 84
2) 120	10) 704	18) 180
3) 24	11) 72	19) 75
4) 60	12) 1,728	20) 224
5) 120	13) 40	21) 144
6) 45	14) 60	22) 153
7) 120	15) 45	23) 288
8) 12	16) 189	24) 48

Divisibility Rule

1) 12	<u>2</u>, <u>3</u>, <u>4</u>, 5, <u>6</u>, 9, 10	13) 18	<u>2</u>, <u>3</u>, 4, 5, <u>6</u>, <u>9</u>, 10
2) 326	<u>2</u>, 3, 4, 5, 6, 9, 10	14) 405	2, <u>3</u>, 4, <u>5</u>, 6, <u>9</u>, 10
3) 748	<u>2</u>, 3, <u>4</u>, 5, 6, 9, 10	15) 945	2, <u>3</u>, 4, <u>5</u>, 6, <u>9</u>, 10
4) 81	2, <u>3</u>, 4, 5, 6, <u>9</u>, 10	16) 186	<u>2</u>, <u>3</u>, 4, 5, <u>6</u>, 9, 10
5) 891	2, <u>3</u>, 4, 5, 6, <u>9</u>, 10	17) 640	<u>2</u>, 3, <u>4</u>, <u>5</u>, 6, 9, <u>10</u>
6) 345	2, <u>3</u>, 4, <u>5</u>, 6, 9, 10	18) 150	<u>2</u>, <u>3</u>, 4, <u>5</u>, 6, 9, <u>10</u>
7) 75	2, <u>3</u>, 4, <u>5</u>, 6, 9, 10	19) 350	<u>2</u>, 3, 4, <u>5</u>, 6, 9, <u>10</u>
8) 450	<u>2</u>, <u>3</u>, 4, <u>5</u>, <u>6</u>, <u>9</u>, <u>10</u>	20) 4,520	<u>2</u>, 3, <u>4</u>, <u>5</u>, 6, 9, <u>10</u>
9) 1,325	2, 3, 4, <u>5</u>, 6, 9, 10	21) 990	<u>2</u>, <u>3</u>, 4, <u>5</u>, <u>6</u>, <u>9</u>, <u>10</u>
10) 78	<u>2</u>, <u>3</u>, 4, 5, <u>6</u>, 9, 10	22) 368	<u>2</u>, 3, <u>4</u>, 5, 6, 9, 10
11) 772	<u>2</u>, 3, <u>4</u>, 5, 6, 9, 10	23) 208	<u>2</u>, 3, <u>4</u>, 5, 6, 9, 10
12) 162	<u>2</u>, <u>3</u>, 4, 5, <u>6</u>, <u>9</u>, 10	24) 500	<u>2</u>, 3, <u>4</u>, <u>5</u>, 6, 9, <u>10</u>

Chapter 2:
Fractions

Adding Fractions – Like Denominator

Find each sum.

1) $\dfrac{1}{4} + \dfrac{2}{4} =$

2) $\dfrac{2}{5} + \dfrac{1}{5} =$

3) $\dfrac{1}{8} + \dfrac{2}{8} =$

4) $\dfrac{4}{11} + \dfrac{1}{11} =$

5) $\dfrac{4}{21} + \dfrac{1}{21} =$

6) $\dfrac{5}{49} + \dfrac{6}{49} =$

7) $\dfrac{2}{7} + \dfrac{11}{7} =$

8) $\dfrac{1}{15} + \dfrac{3}{15} =$

9) $\dfrac{3}{19} + \dfrac{6}{19} =$

10) $\dfrac{1}{13} + \dfrac{1}{13} =$

11) $\dfrac{1}{5} + \dfrac{1}{5} =$

12) $\dfrac{4}{17} + \dfrac{6}{17} =$

13) $\dfrac{2}{20} + \dfrac{17}{20} =$

14) $\dfrac{4}{25} + \dfrac{7}{25} =$

15) $\dfrac{6}{14} + \dfrac{3}{14} =$

16) $\dfrac{12}{30} + \dfrac{5}{30} =$

17) $\dfrac{1}{9} + \dfrac{1}{9} =$

18) $\dfrac{29}{5} + \dfrac{3}{5} =$

19) $\dfrac{18}{6} + \dfrac{5}{6} =$

20) $\dfrac{25}{37} + \dfrac{11}{37} =$

Adding Fractions – Unlike Denominator

Add the fractions and simplify the answers.

1) $\frac{1}{3} + \frac{1}{2} =$

2) $\frac{2}{7} + \frac{2}{3} =$

3) $\frac{3}{6} + \frac{1}{5} =$

4) $\frac{5}{13} + \frac{2}{4} =$

5) $\frac{3}{15} + \frac{2}{5} =$

6) $\frac{16}{56} + \frac{3}{16} =$

7) $\frac{3}{7} + \frac{2}{5} =$

8) $\frac{4}{12} + \frac{2}{5} =$

9) $\frac{6}{13} + \frac{3}{7} =$

10) $\frac{3}{8} + \frac{2}{5} =$

11) $\frac{1}{16} + \frac{4}{6} =$

12) $\frac{5}{24} + \frac{2}{3} =$

13) $\frac{3}{36} + \frac{5}{4} =$

14) $\frac{1}{25} + \frac{2}{5} =$

15) $\frac{7}{49} + \frac{3}{7} =$

16) $\frac{7}{12} + \frac{5}{6} =$

17) $\frac{3}{9} + \frac{2}{5} =$

18) $\frac{3}{45} + \frac{1}{5} =$

19) $\frac{3}{18} + \frac{7}{4} =$

20) $\frac{3}{10} + \frac{1}{4} =$

21) $\frac{3}{64} + \frac{1}{8} =$

22) $\frac{6}{14} + \frac{1}{3} =$

23) $\frac{2}{81} + \frac{1}{3} =$

24) $\frac{6}{15} + \frac{1}{3} =$

Subtracting Fractions – Like Denominator

Find the difference.

1) $\frac{5}{3} - \frac{2}{3} =$

2) $\frac{5}{8} - \frac{3}{8} =$

3) $\frac{11}{14} - \frac{8}{14} =$

4) $\frac{13}{3} - \frac{7}{3} =$

5) $\frac{15}{17} - \frac{13}{17} =$

6) $\frac{18}{33} - \frac{10}{33} =$

7) $\frac{8}{25} - \frac{2}{25} =$

8) $\frac{17}{27} - \frac{2}{27} =$

9) $\frac{7}{10} - \frac{3}{10} =$

10) $\frac{24}{35} - \frac{4}{35} =$

11) $\frac{11}{5} - \frac{3}{5} =$

12) $\frac{28}{38} - \frac{18}{38} =$

13) $\frac{5}{6} - \frac{1}{6} =$

14) $\frac{22}{43} - \frac{11}{43} =$

15) $\frac{4}{7} - \frac{3}{7} =$

16) $\frac{18}{29} - \frac{15}{29} =$

17) $\frac{4}{5} - \frac{3}{5} =$

18) $\frac{42}{53} - \frac{38}{53} =$

19) $\frac{8}{31} - \frac{3}{31} =$

20) $\frac{32}{39} - \frac{30}{39} =$

21) $\frac{9}{26} - \frac{5}{26} =$

22) $\frac{31}{46} - \frac{27}{46} =$

23) $\frac{25}{48} - \frac{19}{48} =$

24) $\frac{39}{65} - \frac{27}{65} =$

Subtracting Fractions – Unlike Denominator

Solve each problem.

1) $\frac{1}{2} - \frac{1}{3} =$

2) $\frac{5}{8} - \frac{2}{5} =$

3) $\frac{5}{6} - \frac{2}{7} =$

4) $\frac{3}{5} - \frac{1}{10} =$

5) $\frac{3}{5} - \frac{5}{12} =$

6) $\frac{5}{8} - \frac{5}{16} =$

7) $\frac{2}{25} - \frac{1}{15} =$

8) $\frac{3}{4} - \frac{13}{18} =$

9) $\frac{8}{5} - \frac{7}{6} =$

10) $\frac{5}{6} - \frac{2}{24} =$

11) $\frac{3}{4} - \frac{5}{36} =$

12) $\frac{1}{5} - \frac{2}{25} =$

13) $\frac{7}{6} - \frac{3}{18} =$

14) $\frac{7}{6} - \frac{5}{12} =$

15) $\frac{3}{5} - \frac{2}{9} =$

16) $\frac{3}{5} - \frac{1}{45} =$

17) $\frac{5}{32} - \frac{5}{48} =$

18) $\frac{2}{3} - \frac{2}{7} =$

19) $\frac{3}{5} - \frac{1}{6} =$

20) $\frac{3}{4} - \frac{5}{13} =$

Converting Mix Numbers

Convert the following mixed numbers into improper fractions.

1) $2\frac{3}{4} =$

2) $4\frac{12}{65} =$

3) $9\frac{3}{7} =$

4) $3\frac{5}{6} =$

5) $6\frac{6}{7} =$

6) $2\frac{10}{24} =$

7) $6\frac{7}{12} =$

8) $2\frac{12}{13} =$

9) $2\frac{12}{10} =$

10) $8\frac{6}{7} =$

11) $6\frac{1}{2} =$

12) $5\frac{14}{16} =$

13) $4\frac{8}{7} =$

14) $2\frac{9}{12} =$

15) $8\frac{3}{5} =$

16) $3\frac{4}{12} =$

17) $6\frac{3}{7} =$

18) $2\frac{1}{15} =$

19) $3\frac{7}{15} =$

20) $4\frac{3}{4} =$

21) $3\frac{5}{9} =$

22) $2\frac{11}{5} =$

23) $5\frac{13}{3} =$

24) $11\frac{7}{13} =$

Converting improper Fractions

Convert the following improper fractions into mixed numbers

1) $\frac{67}{12} =$

2) $\frac{75}{63} =$

3) $\frac{19}{15} =$

4) $\frac{58}{45} =$

5) $\frac{85}{26} =$

6) $\frac{271}{52} =$

7) $\frac{84}{63} =$

8) $\frac{41}{5} =$

9) $\frac{16}{15} =$

10) $\frac{11}{2} =$

11) $\frac{35}{4} =$

12) $\frac{120}{95} =$

13) $\frac{120}{54} =$

14) $\frac{28}{8} =$

15) $\frac{83}{11} =$

16) $\frac{31}{3} =$

17) $\frac{101}{8} =$

18) $\frac{51}{48} =$

19) $\frac{28}{9} =$

20) $\frac{8}{7} =$

21) $\frac{7}{2} =$

22) $\frac{43}{10} =$

23) $\frac{32}{24} =$

24) $\frac{78}{7} =$

Addition Mix Numbers

Add the following fractions.

1) $2\frac{1}{3} + 3\frac{1}{3} =$

2) $6\frac{3}{4} + 2\frac{1}{4} =$

3) $1\frac{1}{7} + 2\frac{2}{7} =$

4) $3\frac{1}{6} + 2\frac{3}{2} =$

5) $3\frac{4}{12} + 3\frac{3}{10} =$

6) $4\frac{1}{7} + 1\frac{1}{2} =$

7) $1\frac{2}{21} + 1\frac{2}{24} =$

8) $3\frac{2}{5} + 1\frac{3}{2} =$

9) $2\frac{3}{5} + 2\frac{1}{5} =$

10) $2\frac{4}{5} + 1\frac{3}{5} =$

11) $3\frac{2}{3} + 1\frac{3}{4} =$

12) $4\frac{1}{6} + 1\frac{3}{7} =$

13) $4\frac{1}{2} + 1\frac{3}{2} =$

14) $5\frac{3}{8} + 2\frac{1}{3} =$

15) $2\frac{3}{4} + 3\frac{1}{3} =$

16) $3\frac{1}{4} + 2\frac{3}{5} =$

17) $2\frac{3}{4} + 8\frac{2}{5} =$

18) $1\frac{3}{4} + 1\frac{1}{2} =$

19) $2\frac{3}{4} + 1\frac{1}{7} =$

20) $1\frac{2}{3} + 1\frac{3}{4} =$

21) $3\frac{1}{6} + 2\frac{1}{4} =$

22) $8\frac{2}{5} + 2\frac{3}{4} =$

23) $4\frac{2}{3} + 5\frac{1}{7} =$

24) $2\frac{1}{3} + 3\frac{2}{5} =$

Subtracting Mix Numbers

Subtract the following fractions.

1) $4\frac{1}{2} - 3\frac{1}{2} =$

2) $3\frac{3}{7} - 3\frac{1}{7} =$

3) $6\frac{3}{5} - 5\frac{1}{5} =$

4) $3\frac{1}{3} - 2\frac{1}{2} =$

5) $4\frac{1}{5} - 3\frac{1}{2} =$

6) $9\frac{1}{3} - 5\frac{2}{3} =$

7) $5\frac{5}{10} - 1\frac{6}{10} =$

8) $7\frac{4}{9} - 5\frac{8}{9} =$

9) $6\frac{2}{11} - 5\frac{5}{11} =$

10) $6\frac{2}{5} - 1\frac{1}{5} =$

11) $9\frac{1}{2} - 5\frac{1}{4} =$

12) $2\frac{5}{8} - 1\frac{3}{8} =$

13) $5\frac{3}{58} - 2\frac{5}{6} =$

14) $5\frac{1}{4} - 3\frac{1}{2} =$

15) $17\frac{1}{8} - 12\frac{3}{8} =$

16) $3\frac{3}{5} - 2\frac{1}{5} =$

17) $2\frac{1}{3} - 1\frac{2}{3} =$

18) $2\frac{1}{6} - 1\frac{2}{3} =$

19) $3\frac{2}{6} - 2\frac{1}{2} =$

20) $2\frac{5}{3} - 2\frac{1}{12} =$

21) $2\frac{9}{10} - 1\frac{1}{5} =$

22) $4\frac{2}{5} - 3\frac{1}{11} =$

23) $2\frac{1}{2} - 1\frac{1}{6} =$

24) $2\frac{3}{10} - 1\frac{4}{10} =$

WWW.MathNotion.com

Simplify Fractions

Reduce these fractions to lowest terms

1) $\frac{24}{16} =$

2) $\frac{18}{27} =$

3) $\frac{12}{15} =$

4) $\frac{36}{48} =$

5) $\frac{9}{27} =$

6) $\frac{15}{35} =$

7) $\frac{28}{49} =$

8) $\frac{80}{100} =$

9) $\frac{9}{81} =$

10) $\frac{25}{10} =$

11) $\frac{24}{32} =$

12) $\frac{20}{60} =$

13) $\frac{24}{40} =$

14) $\frac{3}{12} =$

15) $\frac{14}{49} =$

16) $\frac{52}{78} =$

17) $\frac{96}{36} =$

18) $\frac{48}{180} =$

19) $\frac{12}{32} =$

20) $\frac{88}{77} =$

21) $\frac{160}{320} =$

22) $\frac{24}{124} =$

23) $\frac{144}{36} =$

24) $\frac{120}{480} =$

Multiplying Fractions

Find the product.

1) $\frac{2}{7} \times \frac{3}{8} =$

2) $\frac{4}{25} \times \frac{5}{8} =$

3) $\frac{9}{40} \times \frac{10}{27} =$

4) $\frac{6}{13} \times \frac{22}{33} =$

5) $\frac{9}{12} \times \frac{3}{5} =$

6) $\frac{12}{17} \times \frac{5}{3} =$

7) $\frac{5}{6} \times \frac{6}{5} =$

8) $\frac{35}{89} \times 0 =$

9) $\frac{9}{4} \times \frac{12}{5} =$

10) $\frac{10}{18} \times \frac{3}{5} =$

11) $\frac{36}{25} \times \frac{25}{36} =$

12) $\frac{3}{36} \times \frac{6}{27} =$

13) $\frac{15}{7} \times \frac{3}{5} =$

14) $\frac{6}{7} \times \frac{3}{5} =$

15) $\frac{27}{14} \times \frac{7}{3} =$

16) $\frac{12}{17} \times 0 =$

17) $\frac{7}{11} \times \frac{33}{14} =$

18) $\frac{20}{9} \times \frac{3}{5} =$

19) $\frac{9}{16} \times \frac{4}{81} =$

20) $\frac{4}{23} \times \frac{2}{32} =$

21) $\frac{2}{12} \times \frac{3}{16} =$

22) $\frac{25}{8} \times \frac{2}{125} =$

23) $\frac{9}{16} \times \frac{4}{81} =$

24) $\frac{100}{200} \times \frac{400}{800} =$

Multiplying Mixed Number

Multiply. Reduce to lowest terms.

1) $1\frac{2}{3} \times 1\frac{1}{4} =$

2) $1\frac{2}{5} \times 1\frac{3}{2} =$

3) $1\frac{2}{3} \times 3\frac{1}{8} =$

4) $2\frac{1}{8} \times 1\frac{3}{5} =$

5) $2\frac{2}{3} \times 3\frac{1}{3} =$

6) $2\frac{1}{3} \times 1\frac{2}{3} =$

7) $1\frac{3}{4} \times 2\frac{1}{2} =$

8) $3\frac{2}{3} \times 2\frac{1}{3} =$

9) $2\frac{2}{3} \times 2\frac{1}{2} =$

10) $1\frac{1}{3} \times 1\frac{1}{2} =$

11) $2\frac{3}{4} \times 2\frac{2}{3} =$

12) $3\frac{2}{5} \times 2\frac{4}{7} =$

13) $1\frac{3}{4} \times 2\frac{1}{2} =$

14) $1\frac{1}{2} \times 3\frac{1}{7} =$

15) $1\frac{1}{2} \times 2\frac{1}{5} =$

16) $1\frac{2}{7} \times 2\frac{2}{3} =$

17) $1\frac{2}{3} \times 2\frac{1}{5} =$

18) $1\frac{2}{3} \times 3\frac{2}{5} =$

19) $1\frac{3}{4} \times 2\frac{1}{7} =$

20) $1\frac{1}{3} \times 3\frac{2}{5} =$

21) $1\frac{1}{2} \times 2\frac{1}{6} =$

22) $1\frac{1}{9} \times 1\frac{1}{7} =$

Dividing Fractions

Divide these fractions.

1) $0 \div \frac{1}{5} =$

2) $\frac{6}{12} \div 6 =$

3) $\frac{8}{11} \div \frac{3}{4} =$

4) $\frac{14}{64} \div \frac{2}{8} =$

5) $\frac{3}{19} \div \frac{9}{19} =$

6) $\frac{3}{12} \div \frac{15}{36} =$

7) $9 \div \frac{1}{5} =$

8) $\frac{15}{14} \div \frac{3}{7} =$

9) $\frac{6}{15} \div \frac{1}{14} =$

10) $\frac{2}{13} \div \frac{6}{5} =$

11) $\frac{5}{11} \div \frac{3}{10} =$

12) $\frac{15}{28} \div \frac{3}{7} =$

13) $\frac{7}{16} \div \frac{7}{4} =$

14) $\frac{6}{14} \div \frac{30}{7} =$

15) $\frac{8}{23} \div \frac{2}{23} =$

16) $\frac{9}{32} \div \frac{81}{4} =$

17) $\frac{5}{3} \div \frac{10}{27} =$

18) $8 \div \frac{1}{3} =$

19) $\frac{72}{32} \div \frac{3}{9} =$

20) $\frac{2}{30} \div \frac{8}{5} =$

21) $\frac{2}{9} \div \frac{6}{15} =$

22) $\frac{7}{21} \div \frac{3}{4} =$

Dividing Mixed Number

Divide the following mixed numbers. Cancel and simplify when possible.

1) $2\frac{1}{3} \div 2\frac{1}{2} =$

2) $3\frac{1}{8} \div 2\frac{2}{4} =$

3) $3\frac{1}{2} \div 2\frac{3}{5} =$

4) $2\frac{1}{7} \div 2\frac{1}{2} =$

5) $4\frac{1}{5} \div 2\frac{1}{3} =$

6) $2\frac{5}{9} \div 1\frac{2}{5} =$

7) $2\frac{2}{9} \div 1\frac{1}{2} =$

8) $3\frac{1}{7} \div 2\frac{1}{7} =$

9) $2\frac{1}{9} \div 2\frac{1}{2} =$

10) $3\frac{1}{6} \div 2\frac{2}{3} =$

11) $1\frac{2}{3} \div 5\frac{1}{3} =$

12) $3\frac{1}{9} \div 2\frac{2}{3} =$

13) $3\frac{1}{7} \div 1\frac{1}{11} =$

14) $9\frac{4}{7} \div 4\frac{1}{2} =$

15) $3\frac{3}{4} \div 2\frac{1}{2} =$

16) $2\frac{1}{3} \div 3\frac{2}{5} =$

17) $8\frac{3}{4} \div 2\frac{5}{8} =$

18) $3\frac{1}{3} \div 2\frac{3}{5} =$

19) $3\frac{2}{5} \div 2\frac{1}{2} =$

20) $5\frac{3}{8} \div 2\frac{1}{6} =$

21) $6\frac{1}{2} \div 2\frac{1}{4} =$

22) $4\frac{1}{5} \div 2\frac{1}{7} =$

23) $3\frac{1}{5} \div 2\frac{1}{5} =$

24) $2\frac{1}{7} \div 2\frac{1}{5} =$

Comparing Fractions

Compare the fractions, and write >, < or =

1) $\frac{15}{4}$ ____ $\frac{31}{12}$

2) $\frac{34}{5}$ ____ $\frac{1}{4}$

3) $\frac{3}{6}$ ____ $\frac{7}{5}$

4) $\frac{28}{7}$ ____ $\frac{14}{5}$

5) $\frac{1}{6}$ ____ $\frac{3}{5}$

6) $\frac{11}{7}$ ____ $\frac{15}{9}$

7) $\frac{6}{10}$ ____ $\frac{4}{7}$

8) $\frac{21}{12}$ ____ $\frac{23}{6}$

9) $2\frac{1}{10}$ ____ $5\frac{1}{2}$

10) $4\frac{1}{7}$ ____ $2\frac{1}{6}$

11) $2\frac{1}{3}$ ____ $2\frac{1}{4}$

12) $8\frac{6}{7}$ ____ $8\frac{2}{3}$

13) $1\frac{3}{7}$ ____ $2\frac{5}{3}$

14) $\frac{1}{13}$ ____ $\frac{4}{7}$

15) $\frac{41}{65}$ ____ $\frac{17}{43}$

16) $\frac{65}{200}$ ____ $\frac{65}{92}$

17) $12\frac{1}{2}$ ____ $12\frac{1}{7}$

18) $\frac{1}{2}$ ____ $\frac{1}{4}$

19) $\frac{1}{9}$ ____ $\frac{1}{15}$

20) $\frac{8}{14}$ ____ $\frac{6}{10}$

21) $\frac{5}{25}$ ____ $\frac{8}{56}$

22) $\frac{6}{7}$ ____ $\frac{3}{7}$

23) $1\frac{38}{32}$ ____ $2\frac{3}{16}$

24) $4\frac{18}{5}$ ____ $5\frac{4}{3}$

Answer key Chapter 2

Adding Fractions – Like Denominator

1) $\frac{3}{4}$
2) $\frac{3}{5}$
3) $\frac{3}{8}$
4) $\frac{5}{11}$
5) $\frac{5}{21}$
6) $\frac{11}{49}$
7) $\frac{13}{7}$
8) $\frac{4}{15}$
9) $\frac{9}{19}$
10) $\frac{2}{13}$
11) $\frac{2}{5}$
12) $\frac{10}{17}$
13) $\frac{19}{20}$
14) $\frac{11}{25}$
15) $\frac{9}{14}$
16) $\frac{17}{30}$
17) $\frac{2}{9}$
18) $\frac{32}{5}$
19) $\frac{23}{6}$
20) $\frac{36}{37}$

Adding Fractions – Unlike Denominator

1) $\frac{5}{6}$
2) $\frac{20}{21}$
3) $\frac{7}{10}$
4) $\frac{23}{26}$
5) $\frac{3}{5}$
6) $\frac{53}{112}$
7) $\frac{29}{35}$
8) $\frac{11}{15}$
9) $\frac{81}{91}$
10) $\frac{31}{40}$
11) $\frac{35}{48}$
12) $\frac{7}{8}$
13) $\frac{4}{3}$
14) $\frac{11}{25}$
15) $\frac{4}{7}$
16) $\frac{17}{12}$
17) $\frac{11}{15}$
18) $\frac{4}{15}$
19) $\frac{23}{12}$
20) $\frac{11}{20}$
21) $\frac{11}{64}$
22) $\frac{16}{21}$
23) $\frac{29}{81}$
24) $\frac{11}{15}$

Subtracting Fractions – Like Denominator

1) 1
2) $\frac{1}{4}$
3) $\frac{3}{14}$
4) 2
5) $\frac{2}{17}$
6) $\frac{8}{33}$
7) $\frac{6}{25}$
8) $\frac{5}{9}$
9) $\frac{2}{5}$
10) $\frac{4}{7}$
11) $\frac{8}{5}$
12) $\frac{5}{19}$
13) $\frac{2}{3}$
14) $\frac{11}{43}$
15) $\frac{1}{7}$
16) $\frac{3}{29}$
17) $\frac{1}{5}$
18) $\frac{4}{53}$

SBAC Math Practice Grade 7

19) $\frac{5}{31}$ 21) $\frac{2}{13}$ 23) $\frac{1}{8}$

20) $\frac{2}{39}$ 22) $\frac{2}{23}$ 24) $\frac{12}{65}$

Subtracting Fractions – Unlike Denominator

1) $\frac{1}{6}$ 8) $\frac{1}{36}$ 15) $\frac{17}{45}$

2) $\frac{9}{40}$ 9) $\frac{13}{30}$ 16) $\frac{26}{45}$

3) $\frac{23}{42}$ 10) $\frac{3}{4}$ 17) $\frac{5}{96}$

4) $\frac{1}{2}$ 11) $\frac{11}{18}$ 18) $\frac{8}{21}$

5) $\frac{11}{60}$ 12) $\frac{3}{25}$ 19) $\frac{13}{30}$

6) $\frac{5}{16}$ 13) 1   20) $\frac{19}{52}$

7) $\frac{1}{75}$ 14) $\frac{3}{4}$

Converting Mix Numbers

1) $\frac{11}{4}$ 9) $\frac{32}{10}$ 17) $\frac{45}{7}$

2) $\frac{272}{65}$ 10) $\frac{62}{7}$ 18) $\frac{31}{15}$

3) $\frac{66}{7}$ 11) $\frac{13}{2}$ 19) $\frac{52}{15}$

4) $\frac{23}{6}$ 12) $\frac{94}{16}$ 20) $\frac{19}{4}$

5) $\frac{48}{7}$ 13) $\frac{36}{7}$ 21) $\frac{32}{9}$

6) $\frac{58}{24}$ 14) $\frac{33}{12}$ 22) $\frac{21}{5}$

7) $\frac{79}{12}$ 15) $\frac{43}{5}$ 23) $\frac{28}{3}$

8) $\frac{38}{13}$ 16) $\frac{40}{12}$ 24) $\frac{150}{13}$

Converting improper Fractions

1) $5\frac{7}{12}$ 6) $5\frac{11}{52}$ 11) $8\frac{3}{4}$

2) $1\frac{21}{63}$ 7) $1\frac{21}{63}$ 12) $1\frac{25}{95}$

3) $1\frac{4}{15}$ 8) $8\frac{1}{5}$ 13) $2\frac{12}{54}$

4) $1\frac{13}{45}$ 9) $1\frac{1}{15}$ 14) $3\frac{4}{8}$

5) $3\frac{7}{26}$ 10) $5\frac{1}{2}$ 15) $7\frac{6}{11}$

WWW.MathNotion.com

16) $10\frac{1}{3}$
17) $12\frac{5}{8}$
18) $1\frac{1}{16}$

19) $3\frac{1}{9}$
20) $1\frac{1}{7}$
21) $3\frac{1}{2}$

22) $4\frac{3}{10}$
23) $1\frac{1}{3}$
24) $11\frac{1}{7}$

Adding Mix Numbers

1) $5\frac{2}{3}$
2) 9
3) $3\frac{3}{7}$
4) $6\frac{2}{3}$
5) $6\frac{19}{30}$
6) $5\frac{9}{14}$
7) $2\frac{5}{28}$
8) $5\frac{9}{10}$

9) $4\frac{4}{5}$
10) $4\frac{2}{5}$
11) $5\frac{5}{12}$
12) $5\frac{25}{42}$
13) 7
14) $7\frac{17}{24}$
15) $6\frac{1}{12}$
16) $5\frac{17}{20}$

17) $11\frac{3}{20}$
18) $3\frac{1}{4}$
19) $3\frac{25}{28}$
20) $3\frac{5}{12}$
21) $5\frac{5}{12}$
22) $11\frac{3}{20}$
23) $9\frac{17}{21}$
24) $5\frac{11}{15}$

Subtracting Mix Numbers

1) 1
2) $\frac{2}{7}$
3) $1\frac{2}{5}$
4) $\frac{5}{6}$
5) $\frac{7}{10}$
6) $3\frac{2}{3}$
7) $3\frac{9}{10}$
8) $1\frac{5}{9}$

9) $\frac{8}{11}$
10) $5\frac{1}{5}$
11) $4\frac{1}{4}$
12) $1\frac{1}{4}$
13) $2\frac{19}{87}$
14) $1\frac{3}{4}$
15) $4\frac{3}{4}$
16) $1\frac{2}{5}$

17) $\frac{2}{3}$
18) $\frac{1}{2}$
19) $\frac{5}{6}$
20) $1\frac{7}{12}$
21) $1\frac{7}{10}$
22) $1\frac{17}{55}$
23) $1\frac{1}{3}$
24) $\frac{9}{10}$

Simplify Fractions

1) $\frac{3}{2}$
2) $\frac{2}{3}$
3) $\frac{4}{5}$

4) $\frac{3}{4}$
5) $\frac{1}{3}$
6) $\frac{3}{7}$

7) $\frac{4}{7}$
8) $\frac{4}{5}$
9) $\frac{1}{9}$

SBAC Math Practice Grade 7

10) $\frac{5}{2}$
11) $\frac{3}{4}$
12) $\frac{1}{3}$
13) $\frac{3}{5}$
14) $\frac{1}{4}$

15) $\frac{2}{7}$
16) $\frac{2}{3}$
17) $\frac{8}{3}$
18) $\frac{4}{15}$
19) $\frac{3}{8}$

20) $\frac{8}{7}$
21) $\frac{1}{2}$
22) $\frac{6}{31}$
23) 4
24) $\frac{1}{4}$

Multiplying Fractions

1) $\frac{3}{28}$
2) $\frac{1}{10}$
3) $\frac{1}{12}$
4) $\frac{4}{13}$
5) $\frac{9}{20}$
6) $\frac{20}{17}$
7) 1
8) 0
9) $\frac{27}{5}$

10) $\frac{1}{3}$
11) 1
12) $\frac{1}{54}$
13) $\frac{9}{7}$
14) $\frac{18}{35}$
15) $\frac{9}{2}$
16) 0
17) $\frac{3}{2}$
18) $\frac{4}{3}$

19) $\frac{1}{36}$
20) $\frac{1}{92}$
21) $\frac{1}{32}$
22) $\frac{1}{20}$
23) $\frac{1}{36}$
24) $\frac{1}{4}$

Multiplying Mixed Number

1) $2\frac{1}{12}$
2) $3\frac{1}{2}$
3) $5\frac{5}{24}$
4) $3\frac{2}{5}$
5) $8\frac{8}{9}$
6) $3\frac{8}{9}$
7) $4\frac{3}{8}$
8) $8\frac{5}{9}$

9) $6\frac{2}{3}$
10) 2
11) $7\frac{1}{3}$
12) $8\frac{26}{35}$
13) $4\frac{3}{8}$
14) $4\frac{5}{7}$
15) $3\frac{3}{10}$
16) $3\frac{3}{7}$

17) $3\frac{2}{3}$
18) $5\frac{2}{3}$
19) $3\frac{3}{4}$
20) $4\frac{8}{15}$
21) $3\frac{1}{4}$
22) $1\frac{17}{63}$

Dividing Fractions

1) 0

SBAC Math Practice Grade 7

2) $\frac{1}{12}$
3) $\frac{32}{33}$
4) $\frac{7}{8}$
5) $\frac{1}{3}$
6) $\frac{3}{5}$
7) 45
8) $\frac{5}{2}$

9) $\frac{28}{5}$
10) $\frac{5}{39}$
11) $\frac{50}{33}$
12) $\frac{5}{4}$
13) $\frac{1}{4}$
14) $\frac{1}{10}$
15) 4

16) $\frac{1}{72}$
17) $\frac{9}{2}$
18) 24
19) $\frac{27}{4}$
20) $\frac{1}{24}$
21) $\frac{5}{9}$
22) $\frac{4}{9}$

Dividing Mixed Number

1) $\frac{14}{15}$
2) $1\frac{1}{4}$
3) $1\frac{9}{26}$
4) $\frac{6}{7}$
5) $1\frac{4}{5}$
6) $1\frac{52}{63}$
7) $1\frac{13}{27}$
8) $1\frac{7}{15}$

9) $\frac{38}{45}$
10) $1\frac{3}{16}$
11) $\frac{5}{16}$
12) $1\frac{1}{6}$
13) $2\frac{37}{42}$
14) $2\frac{8}{63}$
15) $1\frac{1}{2}$
16) $\frac{35}{51}$

17) $3\frac{1}{3}$
18) $1\frac{11}{39}$
19) $1\frac{9}{25}$
20) $2\frac{25}{52}$
21) $2\frac{8}{9}$
22) $1\frac{24}{25}$
23) $1\frac{5}{11}$
24) $\frac{75}{77}$

Comparing Fractions

1) >
2) >
3) <
4) >
5) <
6) <

7) >
8) <
9) <
10) >
11) >
12) >

13) <
14) <
15) >
16) <
17) >
18) >

19) >
20) <
21) >
22) >
23) =
24) >

WWW.MathNotion.com

Chapter 3:
Decimal

Round Decimals

Round each number to the correct place value

1) 0.6<u>4</u> =

2) 2.<u>0</u>4 =

3) 6.<u>6</u>23 =

4) 0.<u>3</u>77 =

5) <u>7</u>.707 =

6) 0.0<u>8</u>9 =

7) 6.<u>2</u>4 =

8) 76.7<u>6</u>0 =

9) 1.6<u>2</u>9 =

10) 10.<u>3</u>858 =

11) 1.<u>0</u>9 =

12) 4.<u>2</u>32 =

13) 3.<u>2</u>43 =

14) 6.0<u>5</u>20 =

15) 6<u>3</u>.69 =

16) 3<u>7</u>.32 =

17) 4<u>1</u>9.078 =

18) 512.<u>6</u>55 =

19) 12.3<u>6</u>2 =

20) 6<u>5</u>.65 =

21) 3.2<u>0</u>89 =

22) 37.<u>0</u>73 =

23) 126.<u>5</u>16 =

24) 0.0<u>1</u>22 =

25) 0.07<u>8</u>5 =

26) 5.0<u>1</u>62 =

27) 23.6<u>1</u>33 =

28) 8.0<u>8</u>20 =

Decimals Addition

Add the following.

1) 25.52 + 52.25

2) 0.93 + 0.07

3) 18.96 + 12.87

4) 56.106 + 3.198

5) 6.960 + 5.87

6) 4.148 + 3.231

7) 72.72 + 12.87

8) 56.24 + 23.47

9) 43.06 + 11.87

10) 7.961 + 12.87

11) 18.148 + 12.231

12) 65.98 + 8.37

13) 28.05 + 7.37

14) 125.32 + 3.32

Decimals Subtraction

Subtract the following

1) 8.97 − 2.82

2) 84.02 − 67.57

3) 0.65 − 0.2

4) 9.784 − 7.2

5) 0.784 − 0.05

6) 84.62 − 23.81

7) 121.26 − 78.97

8) 24.36 − 8.38

9) 52.59 − 37.6

10) 5.872 − 0.297

11) 61.43 − 18.8

12) 17.732 − 4.314

13) 23.502 − 2.817

14) 135.35 − 23.56

Decimals Multiplication

Solve.

1) $\begin{array}{r} 2.1 \\ \times\ 2.6 \\ \hline \end{array}$

2) $\begin{array}{r} 8.7 \\ \times\ 5.9 \\ \hline \end{array}$

3) $\begin{array}{r} 7.06 \\ \times\ 2.05 \\ \hline \end{array}$

4) $\begin{array}{r} 67.08 \\ \times\ 10 \\ \hline \end{array}$

5) $\begin{array}{r} 13.08 \\ \times\ 1000 \\ \hline \end{array}$

6) $\begin{array}{r} 32.06 \\ \times\ 7.8 \\ \hline \end{array}$

7) $\begin{array}{r} 26.12 \\ \times\ 12.01 \\ \hline \end{array}$

8) $\begin{array}{r} 4.06 \\ \times\ 7.05 \\ \hline \end{array}$

9) $\begin{array}{r} 18.06 \\ \times\ 0.05 \\ \hline \end{array}$

10) $\begin{array}{r} 21.09 \\ \times\ 9.07 \\ \hline \end{array}$

11) $\begin{array}{r} 14.3 \\ \times\ 15.7 \\ \hline \end{array}$

12) $\begin{array}{r} 5.12 \\ \times\ 0.03 \\ \hline \end{array}$

13) $\begin{array}{r} 8.05 \\ \times\ 0.21 \\ \hline \end{array}$

14) $\begin{array}{r} 12.12 \\ \times\ 5.03 \\ \hline \end{array}$

Decimal Division

Dividing Decimals.

1) $7 \div 1{,}000 =$

2) $3 \div 10 =$

3) $2.6 \div 1{,}000 =$

4) $0.01 \div 100 =$

5) $7 \div 49 =$

6) $2 \div 82 =$

7) $3 \div 48 =$

8) $8 \div 120 =$

9) $8 \div 100 =$

10) $0.8 \div 0.72 =$

11) $0.7 \div 0.07 =$

12) $0.9 \div 0.36 =$

13) $0.5 \div 0.35 =$

14) $0.6 \div 0.06 =$

15) $2.07 \div 10 =$

16) $7.6 \div 100 =$

17) $7.38 \div 1{,}000 =$

18) $15.6 \div 4.5 =$

19) $45.2 \div 5 =$

20) $0.3 \div 0.03 =$

21) $8.05 \div 2.5 =$

22) $0.05 \div 0.20 =$

23) $0.7 \div 4.4 =$

24) $0.08 \div 50 =$

25) $4.16 \div 0.8 =$

26) $0.08 \div 384 =$

Comparing Decimals

Write the Correct Comparison Symbol (>, < or =)

1) 1.15 ____ 2.15

2) 0.4 ____ 0.385

3) 12.5 ____ 12.500

4) 4.05 ____ 4.50

5) 0.511 ____ 0.51

6) 0.623 ____ 0.723

7) 8.76 ____ 8.678

8) 3.0069 ____ 3.069

9) 23.042 ____ 23.034

10) 6.11 ____ 6.08

11) 2.22 ____ 2.222

12) 0.06 ____ 0.55

13) 1.204 ____ 1.25

14) 4.92 ____ 4.0952

15) 0.44 ____ 0.044

16) 17.04 ____ 17.040

17) 0.090 ____ 0.80

18) 20.217 ____ 22.1

19) 0.021 ____ 0.201

20) 21.5 ____ 11.8

21) 3.5 ____ 10.9

22) 0.071 ____ 0.0701

23) 4.021 ____ 0.4021

24) 2.5 ____ 0.255

25) 5.2 ____ 0.255

26) 2.05 ____ 2.0500

27) 6.05 ____ 0.655

28) 1.0501 ____ 1.0510

Convert Fraction to Decimal

Write each as a decimal.

1) $\frac{40}{100} =$

2) $\frac{38}{100} =$

3) $\frac{4}{25} =$

4) $\frac{6}{24} =$

5) $\frac{9}{81} =$

6) $\frac{49}{100} =$

7) $\frac{2}{25} =$

8) $\frac{17}{25} =$

9) $\frac{47}{200} =$

10) $\frac{13}{50} =$

11) $\frac{18}{36} =$

12) $\frac{3}{8} =$

13) $\frac{6}{20} =$

14) $\frac{9}{125} =$

15) $\frac{27}{50} =$

16) $\frac{20}{50} =$

17) $\frac{45}{10} =$

18) $\frac{6}{30} =$

19) $\frac{67}{1,000} =$

20) $\frac{1}{10} =$

21) $\frac{7}{20} =$

22) $\frac{4}{100} =$

Convert Decimal to Percent

Write each as a percent.

1) 0.165 =

2) 0.15 =

3) 1.4 =

4) 0.015 =

5) 0.005 =

6) 0.625 =

7) 0.185 =

8) 0.34 =

9) 0.03 =

10) 0.1 =

11) 0.175 =

12) 4.95 =

13) 2.105 =

14) 0.2 =

15) 1.05 =

16) 0.0275 =

17) 0.0015 =

18) 0.720 =

19) 2.25 =

20) 0.333 =

21) 6.175 =

22) 0.326 =

23) 1.8 =

24) 0.5 =

25) 1.5 =

26) 12.5 =

27) 3.05 =

28) 0.01 =

Convert Fraction to Percent

Write each as a percent.

1) $\frac{1}{5} =$

2) $\frac{5}{4} =$

3) $\frac{8}{16} =$

4) $\frac{19}{22} =$

5) $\frac{14}{20} =$

6) $\frac{13}{50} =$

7) $\frac{7}{9} =$

8) $\frac{13}{20} =$

9) $\frac{5}{100} =$

10) $\frac{8}{20} =$

11) $\frac{3}{25} =$

12) $\frac{14}{100} =$

13) $\frac{48}{50} =$

14) $\frac{32}{50} =$

15) $\frac{19}{28} =$

16) $\frac{3}{33} =$

17) $\frac{24}{44} =$

18) $\frac{23}{28} =$

19) $\frac{24}{84} =$

20) $\frac{5}{50} =$

21) $\frac{25}{625} =$

22) $\frac{480}{240} =$

Answer key Chapter 3

Round Decimals

1) 0.6
2) 2.0
3) 6.6
4) 0.4
5) 8.0
6) 0.09
7) 6.2
8) 76.76
9) 1.63
10) 10.4
11) 1.1
12) 4.2
13) 3.2
14) 6.05
15) 64.0
16) 37.0
17) 420.0
18) 512.7
19) 12.36
20) 66.0
21) 3.21
22) 37.1
23) 126.5
24) 0.01
25) 0.079
26) 5.02
27) 23.61
28) 8.08

Decimals Addition

1) 77.77
2) 1
3) 31.83
4) 59.304
5) 12.83
6) 7.379
7) 85.59
8) 79.71
9) 54.93
10) 20.831
11) 30.379
12) 74.35
13) 35.42
14) 128.64

Decimals Subtraction

1) 6.15
2) 16.45
3) 0.45
4) 2.584
5) 0.734
6) 60.81
7) 42.29
8) 15.98
9) 14.99
10) 5.575
11) 42.63
12) 13.418
13) 20.685
14) 111.79

Decimals Multiplication

1) 5.46
2) 51.33
3) 14.473
4) 670.8
5) 1,3080
6) 250.068
7) 313.7012
8) 28.623
9) 0.903
10) 191.2863
11) 224.51
12) 0.1536
13) 1.6905
14) 60.9636

Decimal Division

1) 0.007
2) 0.3
3) 0.0026

SBAC Math Practice Grade 7

4) 0.0001
5) 0.142…
6) 0.024….
7) 0.0625
8) 0.0666…
9) 0.08
10) 1.111…
11) 10
12) 2.5
13) 1.4285…
14) 10
15) 0.207
16) 0.076
17) 0.00738
18) 3.4666…
19) 9.04
20) 10
21) 3.22
22) 0.25
23) 0.159…
24) 0.0016
25) 5.2
26) 0.0002

Comparing Decimals

1) <
2) >
3) =
4) <
5) >
6) <
7) >
8) <
9) >
10) >
11) <
12) <
13) <
14) >
15) >
16) =
17) <
18) <
19) <
20) >
21) <
22) >
23) >
24) >
25) >
26) =
27) >
28) <

Convert Fraction to Decimal

1) 0.4
2) 0.38
3) 0.16
4) 0.25
5) 0.11
6) 0.49
7) 0.08
8) 0.68
9) 0.235
10) 0.26
11) 0.5
12) 0.375
13) 0.3
14) 0.072
15) 0.54
16) 0.4
17) 4.5
18) 0.2
19) 0.067
20) 0.1
21) 0.35
22) 0.04

Convert Decimal to Percent

1) 16.5%
2) 15%
3) 140%
4) 1.5%
5) 0.5%
6) 62.5%
7) 18.5%
8) 34%
9) 3%

WWW.MathNotion.com

10) 10%
11) 17.5%
12) 495%
13) 210.5%
14) 20%
15) 105%
16) 2.75%

17) 0.15%
18) 72%
19) 225%
20) 33.3%
21) 617.5%
22) 32.6%
23) 180%

24) 50%
25) 150%
26) 1,250%
27) 305%
28) 1%

Convert Fraction to Percent

1) 20%
2) 125%
3) 50%
4) 86.36%
5) 70%
6) 26%
7) 77.8%
8) 65%

9) 5%
10) 40%
11) 12%
12) 14%
13) 96%
14) 64%
15) 67.9%
16) 9.09%

17) 54.5%
18) 82.14%
19) 28.57%
20) 10%
21) 4%
22) 200%

Chapter 4:
Exponent and Radicals

Positive Exponents

Simplify. Your answer should contain only positive exponents.

1) $2^3 =$

2) $5^3 =$

3) $\frac{2x^5y}{xy} =$

4) $(15x3x)^2 =$

5) $(x^3)^2 =$

6) $(\frac{1}{5})^2 =$

7) $0^6 =$

8) $5 \times 5 \times 5 =$

9) $2 \times 2 \times 2 \times 2 \times 2 =$

10) $(3x^2y)^3 =$

11) $10^3 =$

12) $(2x^2y^4)^3 =$

13) $4 \times 10^3 =$

14) $0.5 \times 0.5 \times 0.5 =$

15) $\frac{1}{2} \times \frac{1}{2} \times \frac{1}{2} =$

16) $3^3 =$

17) $(10x^{10}y^3)^2 =$

18) $2^5 =$

19) $x \times x \times x =$

20) $3 \times 3 \times 3 \times 3 \times 3 =$

21) $(3x^2y^3z)^2 =$

22) $7^0 =$

23) $(12x^5y^{-2})^2 =$

24) $(3x^3y^2)^4 =$

WWW.MathNotion.com

Negative Exponents

Simplify. Leave no negative exponents.

1) $3^{-2} =$

2) $7^{-1} =$

3) $(\frac{1}{5})^{-3} =$

4) $10^{-5} =$

5) $1^{-100} =$

6) $4^{-4} =$

7) $(\frac{1}{2})^{-3} =$

8) $-5y^{-3} =$

9) $(\frac{1}{y^{-4}})^{-2} =$

10) $x^{-\frac{3}{2}} =$

11) $\frac{1}{2^{-5}} =$

12) $3^{-4} =$

13) $2^{-3} =$

14) $15^{-1} =$

15) $20^{-2} =$

16) $x^{-4} =$

17) $(x^3)^{-2} =$

18) $x^{-1} \times x^{-1} \times x^{-1} =$

19) $\frac{1}{2} \times \frac{1}{2} =$

20) $10^{-2} =$

21) $10z^{-2} =$

22) $2^{-5} =$

23) $(-\frac{1}{3})^4 =$

24) $6^0 =$

25) $(\frac{1}{x})^{-4} =$

26) $12^{-2} =$

Add and subtract Exponents

Solve each problem.

1) $3^2 + 2^5 =$

2) $x^6 + x^6 =$

3) $3b^2 - 2b^2 =$

4) $3 + 4^3 =$

5) $8 - 4^2 =$

6) $4 + 7^1 =$

7) $2x^3 + 3x^3 =$

8) $10^2 + 3^5 =$

9) $4^5 - 2^4 =$

10) $5^2 - 6^0 =$

11) $1^2 - 3^0 =$

12) $7^1 + 2^3 =$

13) $6^1 - 5^3 =$

14) $3^3 + 3^3 =$

15) $9^2 - 8^2 =$

16) $0^{73} + 0^{54} =$

17) $2^2 - 3^2 =$

18) $7^3 - 7^1 =$

19) $8^2 - 6^2 =$

20) $4^2 + 3^2 =$

21) $2^3 + 4^3 =$

22) $10 + 3^3 =$

23) $6x^5 + 8x^5 =$

24) $8^0 + 4^2 =$

25) $3^2 + 3^2 =$

26) $10^2 + 5^2 =$

27) $(\frac{1}{2})^2 + (\frac{1}{2})^2 =$

28) $9^2 + 3^2 =$

Exponent multiplication

Simplify each of the following

1) $3^6 \times 3^2 =$

2) $9^2 \times 5^0 =$

3) $6^1 \times 7^3 =$

4) $a^{-3} \times a^{-3} =$

5) $y^{-2} \times y^{-2} \times y^{-2} =$

6) $2^4 \times 3^4 \times 2^{-2} \times 3^{-3} =$

7) $5x^2 y^3 \times 8x^3 y^5 =$

8) $(x^2)^3 =$

9) $(x^2 y^3)^4 \times (x^2 y^4)^{-4} =$

10) $6^3 \times 6^2 =$

11) $a^{2b} \times a^0 =$

12) $2^3 \times 2^4 =$

13) $a^m \times a^n =$

14) $a^n \times b^n =$

15) $6^{-2} \times 3^{-2} =$

16) $5^{12} \times 2^{12} =$

17) $(3^5)^4 =$

18) $\left(\frac{1}{5}\right)^3 \times \left(\frac{1}{5}\right)^2 \times \left(\frac{1}{5}\right)^4 =$

19) $\left(\frac{1}{7}\right)^{32} \times 7^{32} =$

20) $(2m)^{\frac{2}{3}} \times (-3m)^{\frac{2}{3}} =$

21) $(x^2 y^3)^{\frac{1}{5}} \times (x^2 y^2)^{\frac{1}{5}} =$

22) $(a^m b^n)^r =$

23) $(3x^2 y^3)^4 =$

24) $(x^{\frac{1}{2}} y^3)^{\frac{-1}{2}} \times (x^2 y^4)^0 =$

25) $6^3 \times 6^4 =$

26) $32^{\frac{1}{4}} \times 32^{\frac{1}{2}} =$

27) $8^4 \times 2^4 =$

28) $(x^3)^0 =$

Exponent division

Simplify. Your answer should contain only positive exponents.

1) $\dfrac{4^3}{4} =$

2) $\dfrac{25x^3}{x} =$

3) $\dfrac{a^m}{a^n} =$

4) $\dfrac{2x^{-5}}{10x^{-3}} =$

5) $\dfrac{81x^8}{9x^3} =$

6) $\dfrac{11x^6}{4x^7} =$

7) $\dfrac{18x^2}{6y^5} =$

8) $\dfrac{35xy^5}{x^5y^2} =$

9) $\dfrac{2x^5}{7x} =$

10) $\dfrac{36x^3y^7}{4x^4} =$

11) $\dfrac{9x^2}{15x^7y^9} =$

12) $\dfrac{yx^4}{5yx^7} =$

13) $\dfrac{14x^2y}{2xy^2} =$

14) $\dfrac{x^{3.25}}{x^{0.25}} =$

15) $\dfrac{5x^3y}{10xy^2} =$

16) $\dfrac{16ab^2r^9}{8a^3b^4} =$

17) $\dfrac{20x^3}{10x^5} =$

18) $\dfrac{16x^3}{4x^6} =$

19) $\dfrac{5^4}{5^2} =$

20) $\dfrac{x}{x^{12}} =$

21) $\dfrac{10^6}{10^2} =$

22) $\dfrac{2xy^4}{8y^2} =$

23) $\dfrac{12x^5y}{144xy^2} =$

24) $\dfrac{42x^6}{7y^8} =$

Scientific Notation

Write each number in scientific notation.

1) 8,100,000 =

2) 50 =

3) 0.0000008 =

4) 254,000 =

5) 0.000225 =

6) 6.5 =

7) 0.00063 =

8) 19,000,000 =

9) 5,000,000 =

10) 85,000,000 =

11) 0.0000036 =

12) 0.00012 =

13) 0.005 =

14) 6,600 =

15) 1,960 =

16) 170,000 =

17) 0.115 =

18) 0.05 =

19) 0.0033 =

20) 20,000 =

21) 23,000 =

22) 0.00000102 =

23) 0.0102 =

24) 1,568 =

25) 32,581 =

26) 12,500 =

27) 12,054 =

28) 60,000 =

Square Roots

Find the square root of each number.

1) $\sqrt{1} =$

2) $\sqrt{4} =$

3) $\sqrt{16} =$

4) $\sqrt{25} =$

5) $\sqrt{49} =$

6) $\sqrt{81} =$

7) $\sqrt{100} =$

8) $\sqrt{144} =$

9) $\sqrt{121} =$

10) $\sqrt{169} =$

11) $\sqrt{9} =$

12) $\sqrt{36} =$

13) $\sqrt{225} =$

14) $\sqrt{196} =$

15) $\sqrt{256} =$

16) $\sqrt{625} =$

17) $\sqrt{289} =$

18) $\sqrt{1,024} =$

19) $\sqrt{484} =$

20) $\sqrt{361} =$

21) $\sqrt{441} =$

22) $\sqrt{841} =$

23) $\sqrt{729} =$

24) $\sqrt{900} =$

25) $\sqrt{400} =$

26) $\sqrt{3,600} =$

27) $\sqrt{4,900} =$

28) $\sqrt{6,400} =$

Simplify Square Roots

Simplify the following.

1) $\sqrt{72} =$

2) $\sqrt{27} =$

3) $\sqrt{28} =$

4) $\sqrt{44} =$

5) $\sqrt{50} =$

6) $\sqrt{40} =$

7) $10\sqrt{125} =$

8) $5\sqrt{600} =$

9) $\sqrt{18} =$

10) $3\sqrt{32} =$

11) $2\sqrt{5} + 8\sqrt{5} =$

12) $\frac{1}{1+\sqrt{2}} =$

13) $\sqrt{20} =$

14) $\frac{5}{2-\sqrt{3}} =$

15) $\sqrt{3} \times \sqrt{12} =$

16) $\frac{\sqrt{400}}{\sqrt{4}} =$

17) $\frac{\sqrt{48}}{\sqrt{16 \times 3}} =$

18) $\sqrt{24y^4} =$

19) $7\sqrt{64a} =$

20) $\sqrt{4+32} + \sqrt{16} =$

21) $\sqrt{90} =$

22) $\sqrt{338} =$

23) $\sqrt{60} =$

24) $\sqrt{75} =$

25) $\sqrt{1,875} =$

26) $\sqrt{32} =$

Answer key Chapter 4

Positive Exponents

1) 8
2) 125
3) $2x^4$
4) $2,025x^4$
5) x^6
6) $\frac{1}{25}$
7) 0
8) 5^3
9) 2^5
10) $27x^6y^3$
11) 1,000
12) $8x^6y^{12}$
13) 4,000
14) 0.5^3
15) $(\frac{1}{2})^3$
16) 27
17) $100x^{20}y^6$
18) 32
19) x^3
20) 3^5
21) $9x^4y^6z^2$
22) 1
23) $\frac{144x^{10}}{y^4}$
24) $81x^{12}y^8$

Negative Exponents

1) $\frac{1}{9}$
2) $\frac{1}{7}$
3) 125
4) $\frac{1}{100,000}$
5) 1
6) $\frac{1}{256}$
7) 8
8) $\frac{-5}{y^3}$
9) y^8
10) $\frac{3}{x^2}$
11) 2^5
12) $\frac{1}{81}$
13) $\frac{1}{8}$
14) $\frac{1}{15}$
15) $\frac{1}{400}$
16) $\frac{1}{x^4}$
17) $\frac{1}{x^6}$
18) $\frac{1}{x^3}$
19) $\frac{1}{2^2}$
20) $\frac{1}{100}$
21) $\frac{10}{z^2}$
22) $\frac{1}{32}$
23) $\frac{1}{81}$
24) 1
25) x^4
26) $\frac{1}{144}$

Add and subtract Exponents

1) 41
2) $2x^6$
3) b^2
4) 67
5) −8
6) 11
7) $5x^3$
8) 343
9) 1,008
10) 24
11) 0
12) 15
13) −119
14) 54
15) 17
16) 0
17) −5
18) 336

SBAC Math Practice Grade 7

19) 28
20) 25
21) 72
22) 37
23) $14x^5$
24) 17
25) 18
26) 125
27) $\frac{1}{2}$
28) 90

Exponent multiplication

1) 3^8
2) 81
3) 2,058
4) a^{-6}
5) y^{-6}
6) $2^2 \times 3^1 = 12$
7) $40x^5y^8$
8) x^6
9) y^{-4}
10) 6^5
11) a^{2b}
12) 2^7
13) a^{m+n}
14) $(ab)^n$
15) 18^{-2}
16) 10^{12}
17) 3^{20}
18) $(\frac{1}{5})^9$
19) 1
20) $(-6m)^{\frac{2}{3}}$
21) $x^{\frac{4}{5}}y$
22) $a^{mr}b^{nr}$
23) $81x^8y^{12}$
24) $x^{\frac{-1}{4}}y^{\frac{-3}{2}}$
25) 6^7
26) $32^{\frac{3}{4}}$
27) $16^4 = 2^{16}$
28) 1

Exponent division

1) 4^2
2) $25x^2$
3) a^{m-n}
4) $\frac{1}{5x^2}$
5) $9x^5$
6) $\frac{11}{4x}$
7) $\frac{3x^2}{y^5}$
8) $\frac{35y^3}{x^4}$
9) $\frac{2x^4}{7}$
10) $\frac{9y^7}{x}$
11) $\frac{3}{5x^5y^9}$
12) $\frac{1}{5x^3}$
13) $\frac{7x}{y}$
14) x^3
15) $\frac{x^2}{2y}$
16) $\frac{2r^9}{a^2b^2}$
17) $\frac{2}{x^2}$
18) $\frac{4}{x^3}$
19) 5^2
20) $\frac{1}{x^{11}}$
21) 10^4
22) $\frac{1}{4}xy^2$
23) $\frac{x^4}{12y}$
24) $\frac{6x^6}{y^8}$

Scientific Notation

1) 81×10^5
2) 5×10^1
3) 8×10^{-7}
4) 2.54×10^5
5) 2.25×10^{-4}
6) 65×10^{-1}
7) 63×10^{-5}
8) 1.9×10^7
9) 5×10^6

SBAC Math Practice Grade 7

10) 8.5×10^7
11) 3.6×10^{-6}
12) 1.2×10^{-4}
13) 5×10^{-3}
14) 6.6×10^3
15) 1.96×10^3
16) 1.7×10^5

17) 1.15×10^{-1}
18) 5×10^{-2}
19) 33×10^{-4}
20) 2×10^4
21) 23×10^3
22) 102×10^{-8}
23) 1.02×10^{-2}

24) 1.568×10^3
25) 32.581×10^3
26) 12.5×10^3
27) 1.2054×10^4
28) 6×10^4

Square Roots

1) 1
2) 2
3) 4
4) 5
5) 7
6) 9
7) 10
8) 12
9) 11
10) 13

11) 3
12) 6
13) 15
14) 14
15) 16
16) 25
17) 17
18) 32
19) 22
20) 19

21) 21
22) 29
23) 27
24) 30
25) 20
26) 60
27) 70
28) 80

Simplify Square Roots

1) $6\sqrt{2}$
2) $3\sqrt{3}$
3) $2\sqrt{7}$
4) $2\sqrt{11}$
5) $5\sqrt{2}$
6) $2\sqrt{10}$
7) $50\sqrt{5}$
8) $50\sqrt{6}$
9) $3\sqrt{2}$

10) $12\sqrt{2}$
11) $10\sqrt{5}$
12) $\sqrt{2} - 1$
13) $2\sqrt{5}$
14) $10 + 5\sqrt{3}$
15) 6
16) 10
17) 1
18) $2y^2\sqrt{6}$

19) $56\sqrt{a}$
20) 10
21) $3\sqrt{10}$
22) $13\sqrt{2}$
23) $2\sqrt{15}$
24) $5\sqrt{3}$
25) $25\sqrt{3}$
26) $4\sqrt{2}$

Chapter 5:
Ratio, Proportion
and Percent

Proportions

Find a missing number in a proportion.

1) $\dfrac{5}{8} = \dfrac{20}{a}$

2) $\dfrac{a}{6} = \dfrac{24}{36}$

3) $\dfrac{14}{42} = \dfrac{a}{3}$

4) $\dfrac{15}{a} = \dfrac{75}{32}$

5) $\dfrac{8}{a} = \dfrac{32}{150}$

6) $\dfrac{\sqrt{16}}{5} = \dfrac{a}{30}$

7) $\dfrac{5}{12} = \dfrac{15}{a}$

8) $\dfrac{6}{12} = \dfrac{a}{33.6}$

9) $\dfrac{8}{a} = \dfrac{3.2}{4}$

10) $\dfrac{1}{16} = \dfrac{3}{a}$

11) $\dfrac{10}{8} = \dfrac{5}{a}$

12) $\dfrac{12}{a} = \dfrac{3}{17}$

13) $\dfrac{2}{7} = \dfrac{a}{10}$

14) $\dfrac{\sqrt{25}}{4} = \dfrac{30}{a}$

15) $\dfrac{12}{a} = \dfrac{13.2}{19.8}$

16) $\dfrac{50}{190} = \dfrac{a}{380}$

17) $\dfrac{32}{100} = \dfrac{a}{52}$

18) $\dfrac{27}{81} = \dfrac{a}{3}$

19) $\dfrac{5}{8} = \dfrac{1}{a}$

20) $\dfrac{5}{3} = \dfrac{35}{a}$

SBAC Math Practice Grade 7

Reduce Ratio

Reduce each ratio to the simplest form.

1) 3: 12 =

2) 4: 24 =

3) 81: 45 =

4) 30: 25 =

5) 24: 240 =

6) 80: 10 =

7) 80: 400 =

8) 5: 180 =

9) 24: 72 =

10) 3.6: 4.2 =

11) 220: 660 =

12) 1.8: 3 =

13) 150: 250 =

14) 40: 60 =

15) 26: 52 =

16) 16: 4 =

17) 100: 25 =

18) 10: 100 =

19) 108: 72 =

20) 130: 165 =

21) 30: 60 =

22) 24: 28 =

23) 10: 150 =

24) 15: 90 =

WWW.MathNotion.com

Percent

Find the Percent of Numbers.

1) 20% of 38 =

2) 42% of 7 =

3) 11% of 11 =

4) 36% of 75 =

5) 5% of 50 =

6) 32% of 14 =

7) 12% of 3 =

8) 9% of 47 =

9) 50% of 52 =

10) 7.5% of 60 =

11) 92% of 12 =

12) 80% of 60 =

13) 12% of 120 =

14) 1% of 310 =

15) 32% of 0 =

16) 62% of 100 =

17) 32% of 44 =

18) 15% of 60 =

19) 5% of 10 =

20) 3% of 7 =

21) 40% of 20 =

22) 70% of 2 =

23) 25% of 20 =

24) 7% of 200 =

25) 50% of 300 =

26) 3% of 6 =

27) 6% of 400 =

28) 9% of 6 =

Discount, Tax and Tip

Find the selling price of each item.

1) Original price of a computer: $250
   Tax: 6%, Selling price: $_____

2) Original price of a laptop: $320
   Tax: 5%, Selling price: $_____

3) Original price of a sofa: $400
   Tax: 7%, Selling price: $_____

4) Original price of a car: $16,500
   Tax: 4.5%, Selling price: $_____

5) Original price of a Table: $300
   Tax: 6%, Selling price: $_____

6) Original price of a house: $450,000
   Tax: 2.5%, Selling price: $_____

7) Original price of a tablet: $200
   Discount: 20%, Selling price: $_____

8) Original price of a chair: $250
   Discount: 15%, Selling price: $_____

9) Original price of a book: $50
   Discount: 35% Selling price: $_____

10) Original price of a cellphone: 600
    Discount: 10% Selling price: $_____

11) Food bill: $32
    Tip: 20%   Price: $_____

12) Food bill: $30
    Tipp: 15%   Price: $_____

13) Food bill: $64
    Tip: 20%   Price: $_____

14) Food bill: $36
    Tipp: 25%   Price: $_____

Find the answer for each word problem.

15) Nicolas hired a moving company. The company charged $200 for its services, and Nicolas gives the movers a 30% tip. How much does Nicolas tip the movers? $_____

16) Mason has lunch at a restaurant and the cost of his meal is $60. Mason wants to leave a 10% tip. What is Mason's total bill including tip? $_____

Percent of Change

Find each percent of change.

1) From 200 to 400. ___ %

2) From 25 ft to 125 ft. ___ %

3) From $50 to $350. ___ %

4) From 40 cm to 160 cm. ___%

5) From 20 to 60. ___ %

6) From 40 to 8. ___ %

7) From 160 to 240. ___ %

8) From 600 to 300. ___ %

9) From 75 to 45. ___ %

10) From 128 to 32. ___ %

Calculate each percent of change word problem.

11) Bob got a raise, and his hourly wage increased from $24 to $30. What is the percent increase? ___ %

12) The price of a pair of shoes increases from $60 to $96. What is the percent increase? ___ %

13) At a coffeeshop, the price of a cup of coffee increased from $2.40 to $2.88. What is the percent increase in the cost of the coffee? ___ %

14) 24cm are cut from a 96 cm board. What is the percent decrease in length? _ %

15) In a class, the number of students has been increased from 108 to 162. What is the percent increase? ___ %

16) The price of gasoline rose from $16.80 to $19.32 in one month. By what percent did the gas price rise? ___ %

17) A shirt was originally priced at $24. It went on sale for $19.20. What was the percent that the shirt was discounted? ___ %

Simple Interest

Determine the simple interest for these loans.

1) $225 at 14% for 2 years. $ _____
2) $2,600 at 8% for 3 years. $ _____
3) $1,300 at 15% for 5 years. $ _____
4) $8,400 at 2.5% for 5 months. $ _____
5) $300 at 2% for 9 months. $ _____
6) $48,000 at 5.5% for 5 years. $ _____
7) $5,200 at 9% for 2 years. $ _____
8) $600 at 5.5% for 4 years. $ _____
9) $800 at 4.5 % for 9 months. $ _____
10) $6,000 at 2.2% for 5 years. $ _____

Calculate each simple interest word problem.

11) A new car, valued at $14,000, depreciates at 4.5% per year. What is the value of the car one year after purchase? $ _____

12) Sara puts $8,000 into an investment yielding 5% annual simple interest; she left the money in for two years. How much interest does Sara get at the end of those two years? $ _____

13) A bank is offering 10.5% simple interest on a savings account. If you deposit $22,500, how much interest will you earn in two years? $ _____

14) $800 interest is earned on a principal of $8,000 at a simple interest rate of 5% interest per year. For how many years was the principal invested? _____

15) In how many years will $1,500 yield an interest of $300 at 5% simple interest? _____

16) Jim invested $6,000 in a bond at a yearly rate of 3.5%. He earned $630 in interest. How long was the money invested? _____

Answer key Chapter 5

Proportions

1) $a = 32$
2) $a = 4$
3) $a = 1$
4) $a = 6.4$
5) $a = 37.5$
6) $a = 24$
7) $a = 36$
8) $a = 16.8$
9) $a = 10$
10) $a = 48$
11) $a = 4$
12) $a = 68$
13) $a = \frac{20}{7}$
14) $a = 24$
15) $a = 18$
16) $a = 100$
17) $a = 16.64$
18) $a = 1$
19) $a = 1.6$
20) $a = 21$

Reduce Ratio

1) $1:4$
2) $1:6$
3) $9:5$
4) $6:5$
5) $1:10$
6) $8:1$
7) $1:5$
8) $1:36$
9) $1:3$
10) $0.6:0.7$
11) $11:33$
12) $0.6:1$
13) $3:5$
14) $2:3$
15) $1:2$
16) $4:1$
17) $4:1$
18) $1:10$
19) $3:2$
20) $26:33$
21) $1:2$
22) $6:7$
23) $1:15$
24) $1:6$

Percent

1) 7.6
2) 2.94
3) 1.21
4) 27
5) 2.5
6) 4.48
7) 0.36
8) 4.23
9) 26
10) 4.5
11) 11.04
12) 48
13) 14.4
14) 3.1
15) 0
16) 62
17) 14.08
18) 9
19) 0.5
20) 0.21
21) 8
22) 1.4
23) 5
24) 14
25) 150
26) 0.18
27) 24
28) 0.54

Discount, Tax and Tip

1) $265.00
2) $336.00
3) $428.00
4) $17,242.50
5) $318.00
6) $461,250
7) $240.00
8) $287.50
9) $67.50
10) $660.00
11) $38.40
12) $34.50
13) $76.80
14) $45.00
15) $60.00
16) $66.00

Percent of Change

1) 100%
2) 400%
3) 600%
4) 300%
5) 200%
6) 80%
7) 50%
8) 50%
9) 40%
10) 75%
11) 25%
12) 60%
13) 20%
14) 25%
15) 50%
16) 15%
17) 20%

Simple Interest

1) $63.00
2) $624.00
3) $975.00
4) $87.50
5) $4.50
6) $13,200.00
7) $936.00
8) $132.00
9) $27.00
10) $660.00
11) $13,370.00
12) $800.00
13) $4725.00
14) 2 *years*
15) 4 *years*
16) 3 *years*

Chapter 6:
Measurement

Reference Measurement

LENGTH	
Customary	**Metric**
1 mile (mi) = 1,760 yards (yd)	1 kilometer (km) = 1,000 meters (m)
1 yard (yd) = 3 feet (ft)	1 meter (m) = 100 centimeters (cm)
1 foot (ft) = 12 inches (in.)	1 centimeter(cm) = 10 millimeters(mm)
VOLUME AND CAPACITY	
Customary	**Metric**
1 gallon (gal) = 4 quarts (qt)	1 liter (L) = 1,000 milliliters (mL)
1 quart (qt) = 2 pints (pt.)	
1 pint (pt.) = 2 cups (c)	
1 cup (c) = 8 fluid ounces (Fl oz)	
WEIGHT AND MASS	
Customary	**Metric**
1 ton (T) = 2,000 pounds (lb.)	1 kilogram (kg) = 1,000 grams (g)
1 pound (lb.) = 16 ounces (oz)	1 gram (g) = 1,000 milligrams (mg)
Time	
1 year = 12 months	
1 year = 52 weeks	
1 week = 7 days	
1 day = 24 hours	
1 hour = 60 minutes	
1 minute = 60 seconds	

Metric Length Measurement

Convert to the units.

1) 5×10^4 mm = _____ cm

2) 0.4 m = _____ mm

3) 0.06 m = _____ cm

4) 1.2 km = _____ m

5) 8,000 mm = _____ m

6) 4,700 cm = _____ m

7) 4.5 m = _____ cm

8) 7×10^3 mm = _____ cm

9) 9×10^6 mm = _____ m

10) 2 km = _____ mm

11) 0.3 km = _____ m

12) 0.05 m = _____ cm

13) 4×10^4 m = _____ km

14) 6×10^7 m = _____ km

Customary Length Measurement

Convert to the units.

1) 20 ft = _____ in

2) 2.5 ft = _____ in

3) 5.6 yd = _____ ft

4) 0.4 yd = _____ ft

5) 9×10^{-1} yd = _____ in

6) 2 mi = _____ in

7) 18×10^3 in = _____ yd

8) 21.6 in = _____ yd

9) 6,160 yd = _____ mi

10) 28 yd = _____ in

11) 0.03 mi = _____ yd

12) 99×10^3 ft = _____ mi

13) 4.8 in = _____ ft

14) 42 yd = _____ feet

15) 0.72 in = _____ ft

16) 0.2 mi = _____ ft

Metric Capacity Measurement

Convert the following measurements.

1) 60 l = _____ ml

2) 0.7 l = _____ ml

3) 2.8 l = _____ ml

4) 0.06 l = _____ ml

5) 22.5 l = _____ ml

6) 0.9 l = _____ ml

7) 6×10^6 ml = _____ l

8) 22×10^5 ml = _____ l

9) 112×10^2 ml = _____ l

10) 11,000 ml = _____ l

11) 57,800 ml = _____ l

12) 0.3×10^5 ml = _____ l

Customary Capacity Measurement

Convert the following measurements.

1) 1.5 gal = _____ qt.

2) 4.5 gal = _____ pt.

3) 0.5 gal = _____ c.

4) 18 pt. = _____ c

5) 12 c = _____ fl oz

6) 8.15 qt = _____ pt.

7) 0.08 qt = _____ c

8) 42 pt. = _____ c

9) 8×10^4 c = _____ gal

10) 256 pt. = _____ gal

11) 484 qt = _____ gal

12) 25.8 pt. = _____ qt

13) 7×10^3 c = _____ qt

14) 98.8 c = _____ pt.

15) 0.164 qt = _____ gal

16) 1,256 pt. = _____ qt

17) 23 gal = _____ pt.

18) 0.01 qt = _____ c

19) 800 c = _____ gal

20) 64.16 fl oz = _____ c

Metric Weight and Mass Measurement

Convert.

1) 0.8 kg = _____ g

2) 5.6 kg = _____ g

3) 2×10^{-4} kg = _____ g

4) 1.04 kg = _____ g

5) 44.8 kg = _____ g

6) 13.12 kg = _____ g

7) 0.072 kg = _____ g

8) 21×10^5 g = _____ kg

9) 15×10^6 g = _____ kg

10) 0.04×10^8 g = _____ kg

11) 17,400 g = _____ kg

12) 98×10^2 g = _____ kg

13) 5,400,000 g = _____ kg

14) 325×10^4 g = _____ kg

Customary Weight and Mass Measurement

Convert.

1) 24×10^4 lb. = _____ T

2) 0.32×10^5 lb. = _____ T

3) 190,000 lb. = _____ T

4) 2,800 lb. = _____ T

5) 0.35 lb. = _____ oz

6) 2.8 lb. = _____ oz

7) 0.05 lb. = _____ oz

8) 4 T = _____ lb.

9) 7×10^{-4} T = _____ lb.

10) 38×10^{-5} T = _____ lb.

11) 0.6 T = _____ lb.

12) 0.003 T = _____ oz

13) 0.015 T = _____ oz

14) 196.8 oz = _____ lb.

Unit of Measurements

Use the given ratios to convert the measuring units. If necessary, round the answers to three decimal digits.

1) Use $1 = \frac{1.6093 km}{1 mi}$ and convert 6.02 miles to kilometers

   6.02 mi = _____

2) Use $1 = \frac{1.6093 km}{1 mi}$ and convert 4.15 miles to kilometers

   4.15 mi = _____

3) Use $1 = \frac{1 qt}{0.946 L}$ and convert 6 liters to quarts

   6 L = _____

4) Use $1 = \frac{1 qt}{0.946 L}$ and convert 8 liters to quarts

   8 L = _____

5) Use $1 = \frac{1.6093 km}{1 mi}$ and convert 5.06 miles to kilometers

   5.06 mi = _____

6) Use $1 = \frac{1.6093 km}{1 mi}$ and convert 8.1 miles to kilometers

   8.1 mi = _____

7) Use $1 = \frac{1 qt}{0.946 L}$ and convert 5 liters to quarts

   5 L = _____

SBAC Math Practice Grade 7

Temperature

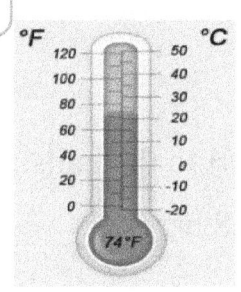

Convert Fahrenheit into Celsius.

1) 176°F = ___ °C

2) 113°F = ___ °C

3) 122°F = ___ °C

4) −13°F = ___ °C

5) 131°F = ___ °C

6) 136.4°F = ___ °C

7) 14°F = ___ °C

8) 149°F = ___ °C

9) 158°F = ___ °C

10) 167°F = ___ °C

11) 77°F = ___ °C

12) 86°F = ___ °C

Convert Celsius into Fahrenheit.

13) 85°C = ___ °F

14) 150°C = ___ °F

15) 83°C = ___ °F

16) 20°C = ___ °F

17) 5°C = ___ °F

18) −5°C = ___ °F

19) 0°C = ___ °F

20) 30°C = ___ °F

21) 90°C = ___ °F

22) 72°C = ___ °F

23) 38°C = ___ °F

24) 35°C = ___ °F

WWW.MathNotion.com

Time

Convert to the units.

1) 28 hr. = _____ min

2) 15 year = _____ week

3) 0.5 hr. = _____ sec

4) 8.5 min = _____ sec

5) 6×10^4 min = _____ hr

6) 1,095 day = _____ year

7) 2 year = _____ hr.

8) 42 day = _____ hr

9) 2 day = _____ min

10) 480 min = _____ hr

11) 28.5 year = _____ month

12) 12,600 sec = _____ min

13) 216 hr = _____ day

14) 15 weeks = _____ day

How much time has passed?

1) From 3:35 A.M. to 6:45 A.M.: ____ hours and ____ minutes.

2) From 2:30 A.M. to 7:15 A.M.: ____ hours and ____ minutes.

3) It's 6:20 P.M. What time was 3 hours ago? _____ O'clock

4) 4:15 A.M to 7:35 AM: _____ hours and _____ minutes.

5) 1:45 A.M to 4:20 AM: _____ hours and _____ minutes.

6) 9:00 A.M. to 10:05 AM. = _____ hour(s) and _____ minutes.

7) 10:35 A.M. to 3:05 PM. = _____ hour(s) and _____ minutes

8) 5:12 A.M. to 5:48 A.M. = _____ minutes

9) 8:08 A.M. to 8:45 A.M. = _____ minutes

Answers of Worksheets – Chapter 6

Metric length

1) 5,000 cm
2) 400 mm
3) 6 cm
4) 1,200 m
5) 8 m
6) 47 m
7) 450 cm
8) 700 cm
9) 9,000 m
10) 2,000,000 mm
11) 300 m
12) 5 cm
13) 40 km
14) 60,000 km

Customary Length

1) 240
2) 30
3) 16.8
4) 1.2
5) 32.4
6) 126,720
7) 500
8) 0.6
9) 3.5
10) 1,008
11) 52.8
12) 18.75
13) 0.4
14) 126
15) 0.06
16) 1,056

Metric Capacity

1) 60,000 ml
2) 700 ml
3) 2,800 ml
4) 60 ml
5) 22,500 ml
6) 900 ml
7) 6,000 ml
8) 2,200 ml
9) 11.2 ml
10) 11 L
11) 57.8 L
12) 30 L

Customary Capacity

1) 6 qt
2) 36 pt.
3) 8 c
4) 36 c
5) 96 fl oz
6) 16.3 pt.
7) 0.32 c
8) 84 c
9) 5,000 gal
10) 32 gal
11) 121 gal
12) 12.9 qt
13) 1,750 qt
14) 49.4 pt.
15) 0.041 gal
16) 628 qt
17) 184 pt.
18) 0.04 c
19) 50 gal
20) 8.02 c

Metric Weight and Mass

1) 800 g
2) 5,600 g
3) 0.2 g
4) 1,040 g
5) 44,800 g
6) 13,120 g
7) 72 g
8) 2,100 kg
9) 15,000 kg
10) 4,000 kg
11) 17.4 kg
12) 9.8 kg
13) 5,400 kg
14) 3,250 kg

Customary Weight and Mass

1) 120 T
2) 16 T
3) 95 T
4) 1.4 T
5) 5.6 oz
6) 44.8 oz
7) 0.8 oz
8) 8,000 lb.
9) 1.4 lb.
10) 0.76 lb.
11) 1,200 lb.
12) 96 oz
13) 480 oz
14) 12.3 lb

Unit of measurements

1) 9.688km
2) 6.679km
3) 6.342 qt
4) 8.457qt
5) 8.143km
6) 13.035km
7) 5.285qt

Temperature

1) 80°C
2) 45°C
3) 50°C
4) −25°C
5) 55°C
6) 58°C
7) −10°C
8) 65°C
9) 70°C
10) 75°C
11) 25°C
12) 30°C
13) 185°F
14) 302°F
15) 181.4°F
16) 68°F
17) 41°F
18) 23°F
19) 32°F
20) 86°F
21) 194°F
22) 161.6°F
23) 100.4°F
24) 95°F

Time - Convert

1) 1,680 min
2) 780 weeks
3) 1,800 sec
4) 510 sec
5) 1,000 hr
6) 3 year
7) 17,520 hr
8) 1,008 hr
9) 2,880 min
10) 8 hr
11) 342 months
12) 210 min
13) 9 days
14) 105 days

Time - Gap

1) 3:10
2) 4:45
3) 3:20 P.M.
4) 3:20
5) 2:35
6) 1:05
7) 4:30
8) 36 minutes
9) 37 minutes

Chapter 7:
Linear Functions

Relation and Functions

Determine whether each relation is a function. Then state the domain and range of each relation.

1)
Function:
....
Domain:
....
Range:
....

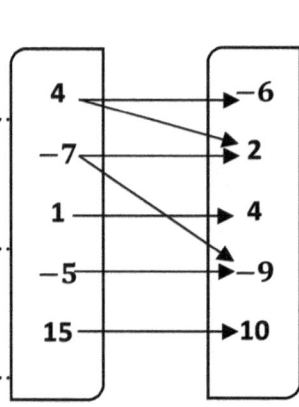

2)
Function:
....
Domain:
....
Range:
....

x	y
4	5
2	3
−6	−8
6	−8
−11	2

3)
Function:
....
Domain:
....
Range:
....

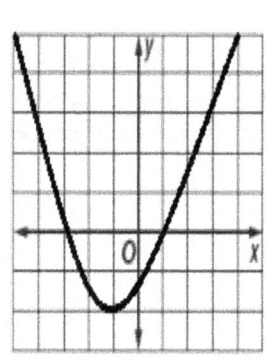

4) $\{(2, -2), (7, -6), (9, 9), (8, 1), (7, 4)\}$

Function:
....
Domain:
....
Range:
....

5)
Function:
....
Domain:
....
Range:
....

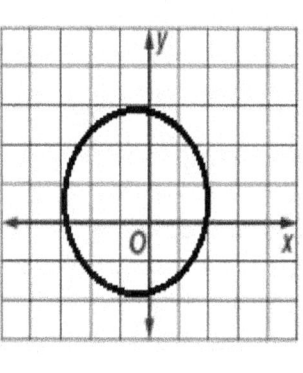

6)
Function:
....
Domain:
....
Range:
....

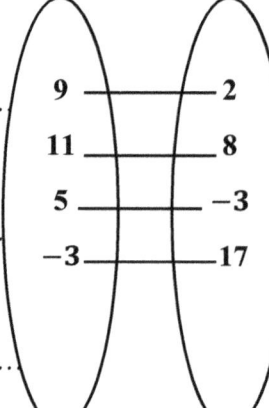

SBAC Math Practice Grade 7

Slope form

Write the slope-intercept form of the equation of each line.

1) $4x + 5y = 15$

2) $4x + 12y = 3$

3) $7x + y = -9$

4) $-7x + 11y = 5$

5) $5x - 4y = 7$

6) $-21x + 3y = 6$

7) $2x + y = 0$

8) $5x - 7y = -9$

9) $-13.5x + 27y = 54$

10) $-3x + \frac{2}{3}y = 18$

11) $10x + y = -120$

12) $3x = -36y - 27$

13) $1.5x = 3y + 3$

14) $5x = -\frac{5}{4}y + 25$

Slope and Y-Intercept

Find the slope and y-intercept of each equation.

1) $y = \frac{1}{5}x + 4$

2) $y = 7x + 8$

3) $x - 3y = 9$

4) $y = 5x + 21$

5) $y = 9$

6) $y = -2x + 3$

7) $x = -15$

8) $y = 7x$

9) $y - 3 = 4(x + 1)$

10) $x = -\frac{5}{8}y - \frac{1}{3}$

WWW.MathNotion.com

Slope and One Point

Find a Point-Slope equation for a line containing the given point and having the given slope.

1) $m = -2, (1, -1)$

2) $m = 3, (1, 2)$

3) $m = -2, (-1, -5)$

4) $m = 1, (3, 2)$

5) $m = 5, (2, 4)$

6) $m = \frac{3}{2}, (4, 5)$

7) $m = 0, (-4, -5)$

8) $m = 2, (1, -3)$

9) $m = 1, (0, 3)$

10) $m = \frac{3}{4}, (-2, -5)$

11) $m = -3, (1, -1)$

12) $m = -2, (2, -1)$

13) $m = 5, (1, 0)$

14) $m =$ undefined, $(8, -8)$

15) $m = -\frac{1}{8}, (8, 4)$

16) $m = \frac{1}{4}, (3, 2)$

17) $m = -8, (2, 4)$

18) $m = 6, (-2, -4)$

19) $m = \frac{1}{3}, (3, 1)$

20) $m = \frac{-4}{9}, (0, -3)$

21) $m = \frac{1}{4}, (4, 4)$

22) $m = -5, (0, -1)$

23) $m = 0, (0.9, -3)$

24) $m = -\frac{5}{7}, (7, -1)$

25) $m = 0, (-4, 8)$

26) $m =$ Undefined, $(-10, -2)$

WWW.MathNotion.com

Slope of Two Points

Write the slope-intercept form of the equation of the line through the given points.

1) $(1, 0), (-1, 5)$

2) $(-1, 3), (5, 6)$

3) $(-5, 1), (-1, 5)$

4) $(2, -3), (-9, 8)$

5) $(5, 0), (3, 1)$

6) $(9, -1), (-1, 9)$

7) $(-5, 3), (-6, 1)$

8) $(-7, -2), (1, 0)$

9) $(-5, -5), (3, 3)$

10) $(-1, 9), (-1, -5)$

11) $(-2, 7), (1, 7)$

12) $(1, -5), (4, -4)$

13) $(6, -9), (-3, 0)$

14) $(1, -4), (7, 4)$

15) $(-9, 5), (-3, -1)$

16) $(9, 5), (5, 1)$

17) $(10, -7), (2, -6)$

18) $(-5, -9), (-7, 2)$

19) $(7, 4), (3, 1)$

20) $(-1, -1), (9, 2)$

21) $(-8, 8), (8, 2)$

22) $(9, 2), (5, 11)$

23) $(8, 2), (9, 3)$

24) $(-2, -5), (-5, -8)$

Equation of Parallel and Perpendicular lines

Write the slope-intercept form of the equation of the line described.

1) Through: $(-5, 2)$, parallel to $y = 2x + 5$

2) Through: $(-4, 1)$, parallel to $y = -3x$

3) Through: $(-10, -2)$, perpendecular to $y = \frac{1}{2}x + 8$

4) Through: $(6, -2)$, parallel to $y = -5x + 13$

5) Through: $(-7, 4)$, parallel to $y = \frac{3}{7}x - 6$

6) Through: $(2, 0)$, perpendecular to $y = -\frac{1}{5}x + 8$

7) Through: $(4, -7)$, perpendecular to $y = -6x - 10$

8) Through: $(-5, 1)$, perpendecular to $y = -\frac{1}{8}x + 3$

9) Through: $(-1, -2)$, parallel to $2y + 4x = 9$

10) Through: $(1, 10)$, parallel to $y = \frac{1}{10}x - 5$

11) Through: $(5, -5)$, parallel to $y = 9$

12) Through: $(7, 2)$, perpendecular to $y = \frac{5}{2}x + 3$

13) Through: $(0, -4)$, perpendecular to $3y - x = 11$

14) Through: $(3, 5)$, parallel to $3y + x = 5\frac{3}{4}$

15) Through: $(1, 1)$, perpendecular to $y = 5x + 12$

16) Through: $(-3, -5)$, parallel to $8y - x = 10$

17) Through: $(-2, -2)$, perpendecular to $y = 4x + \frac{1}{7}$

18) Through: $(-8, 0)$, perpendecular to $5y - 4x - 9 = 0$

Quadratic Equations - Quadratic Formula

Solve each equation with the quadratic formula.

1) $3x^2 + 6x - 24 = 0$

2) $4x^2 - 8x = -4$

3) $\frac{1}{4}x^2 = \frac{9}{4}x - 5$

4) $\frac{2}{3}x^2 + \frac{10}{3}x - 4 = 0$

5) $3x^2 = 27x - 60$

6) $3x^2 - 12x - 26 = 10$

7) $7x^2 = -21x + 280$

8) $3x^2 + 15x - 18 = 0$

9) $x^2 + x - 2 = \frac{1}{4}$

10) $\frac{4}{3}x^2 - \frac{2}{3}x = 3$

11) $x^2 = -3x + 40$

12) $9x^2 - 25 = 8x$

13) $\frac{8}{5}x^2 - \frac{8}{5}x = 3$

14) $6x^2 - 3x - 10 = 8$

15) $\frac{1}{3}x^2 = 3x - \frac{100}{15}$

16) $10x^2 + 30x = 400$

17) $x^2 - 2 = \frac{1}{8}x$

18) $24x^2 + 18x + 15 = 0$

19) $x^2 - \frac{1}{2}x - \frac{13}{2} = 1$

20) $11x^2 + 1 = 5x^2 + 7x$

21) $17x^2 + 14 = x$

22) $10x^2 - 5x - 20 = 10$

Answer key Chapter 7

Relation and Functions

1) No, $D_f = \{4, -7, 1, -5, 15\}$, $R_f = \{-6, 2, 4, -9, 10\}$
2) Yes, $D_f = \{4, 2, -6, 6, -11\}$, $R_f = \{5, 3, -8, 2\}$
3) Yes, $D_f = (-\infty, \infty)$, $R_f = \{-2, \infty)$
4) No, $D_f = \{2, 7, 9, 8, 7\}$, $R_f = \{-2, -6, 9, 1, 4\}$
5) No, $D_f = [-3, 2]$, $R_f = [-2, 3]$
6) Yes, $D_f = \{9, 11, 5, -3\}$, $R_f = \{2, 8, -3, 17\}$

Slope form

1) $y = -\frac{4}{5}x + 3$
2) $y = -\frac{1}{3}x + \frac{1}{4}$
3) $y = -7x - 9$
4) $y = \frac{7}{11}x + \frac{5}{11}$
5) $y = \frac{5}{4}x - \frac{7}{4}$
6) $y = 7x + 2$
7) $y = -2x$
8) $y = \frac{5}{7}x + \frac{9}{7}$
9) $y = 0.5x + 2$
10) $y = 4.5x + 27$
11) $y = -10x - 120$
12) $y = -\frac{1}{12}x - \frac{3}{4}$
13) $y = 0.5x - 1$
14) $y = -4x + 5$

Slope and Y-Intercept

1) $m = \frac{1}{5}, b = 4$
2) $m = 7, b = 8$
3) $m = \frac{1}{3}, b = -3$
4) $m = 5, b = 21$
5) $m = 0, b = 9$
6) $m = -2, b = 3$
7) $m = undefind$, $b: no\ intercept$
8) $m = 7, b = 0$
9) $m = 4, b = 7$
10) $m = -\frac{8}{5}, b = -\frac{1}{3}$

Slope and One Point

1) $y = -2x + 1$
2) $y = 3x - 1$
3) $y = -2x - 7$
4) $y = x - 1$
5) $y = 5x - 6$
6) $y = \frac{3}{2}x - 1$
7) $y = -5$
8) $y = 2x - 5$
9) $y = x + 3$
10) $y = \frac{3}{4}x - \frac{7}{2}$
11) $y = -3x + 2$
12) $y = -2x + 3$
13) $y = 5x$
14) $x = 8$
15) $y = -\frac{1}{8}x + 5$
16) $y = \frac{1}{4}x + \frac{5}{4}$
17) $y = -8x + 20$
18) $y = 6x + 8$
19) $y = \frac{1}{3}x$
20) $y = -\frac{4}{9}x - 3$
21) $y = \frac{1}{4}x + 3$
22) $y = -5x - 1$

SBAC Math Practice Grade 7

23) $y = -3$
24) $y = -\frac{5}{7}x + 4$
25) $y = 8$
26) $x = -10$

Slope of Two Points

1) $y = -\frac{5}{2}x + \frac{5}{2}$
2) $y = \frac{1}{2}x + \frac{7}{2}$
3) $y = x + 6$
4) $y = -x - 1$
5) $y = -\frac{1}{2}x + \frac{5}{2}$
6) $y = -x + 8$
7) $y = 2x + 13$
8) $y = \frac{1}{4}x - \frac{1}{4}$
9) $y = x$
10) $x = -1$
11) $y = 7$
12) $y = \frac{1}{3}x - 5\frac{1}{3}$
13) $y = -x - 3$
14) $y = \frac{4}{3}x - 5\frac{1}{3}$
15) $y = -x - 4$
16) $y = x - 4$
17) $y = -\frac{1}{8}x - 5\frac{3}{4}$
18) $y = -5\frac{1}{2}x - 36\frac{1}{2}$
19) $y = \frac{3}{4}x - 1\frac{1}{4}$
20) $y = \frac{3}{10}x - \frac{7}{10}$
21) $y = -\frac{3}{8}x + 5$
22) $y = -\frac{9}{4}x + 22\frac{1}{4}$
23) $y = x - 6$
24) $y = x - 3$

Equation of Parallel and Perpendicular lines

1) $y = 2x + 12$
2) $y = -3x - 11$
3) $y = -2x - 22$
4) $y = -5x + 28$
5) $y = \frac{3}{7}x + 7$
6) $y = 5x - 10$
7) $y = \frac{1}{6}x - 7\frac{2}{3}$
8) $y = 8x + 41$
9) $y = -2x - 4$
10) $y = \frac{1}{10}x + 9\frac{9}{10}$
11) $y = -5$
12) $y = -\frac{2}{5}x + 4\frac{4}{5}$
13) $y = -3x - 4$
14) $y = -\frac{1}{3}x + 6$
15) $y = -\frac{1}{5}x + 1\frac{1}{5}$
16) $y = \frac{1}{8}x - 4\frac{5}{8}$
17) $y = -\frac{1}{4}x - 2\frac{1}{2}$
18) $y = -\frac{5}{4}x - 10$

Quadratic Equations - Quadratic Formula

1) $\{2, -4\}$
2) $\{1\}$
3) $\{5, 4\}$
4) $\{1, -6\}$
5) $\{5, 4\}$
6) $\{6, -2\}$
7) $\{5, -8\}$
8) $\{1, -6\}$
9) $\{\frac{-1+\sqrt{10}}{2}, \frac{-1-\sqrt{10}}{2}\}$
10) $\{\frac{1+\sqrt{37}}{4}, \frac{1-\sqrt{37}}{4}\}$
11) $\{5, -8\}$
12) $\{\frac{4+\sqrt{241}}{9}, \frac{4-\sqrt{241}}{9}\}$
13) $\{\frac{2+\sqrt{34}}{4}, \frac{2-\sqrt{34}}{4}\}$
14) $\{2, -\frac{3}{2}\}$
15) $\{5, 4\}$
16) $\{5, -8\}$
17) $\{\frac{1+3\sqrt{57}}{16}, \frac{1-3\sqrt{57}}{16}\}$
18) $\{\frac{-3+i\sqrt{31}}{8}, \frac{-3-i\sqrt{31}}{8}\}$
19) $\{3, -\frac{5}{2}\}$
20) $\{1, \frac{1}{6}\}$
21) $\{\frac{1+i\sqrt{951}}{34}, \frac{1-i\sqrt{951}}{34}\}$
22) $\{2, -\frac{3}{2}\}$

WWW.MathNotion.com

Chapter 8:
Equations and Inequality

Distributive and Simplifying Expressions

Simplify each expression.

1) $6x + 2 - 8 =$

2) $-(-4 - 5x) =$

3) $(-3x + 4)(-2) =$

4) $(-2x)(x + 3) =$

5) $-2x + x^2 + 4x^2 =$

6) $7y + 7x + 8y - 5x =$

7) $-3x + 3y + 14x - 9y =$

8) $-2x - 5 + 8x + \frac{16}{4} =$

9) $5 - 8(x - 2) =$

10) $-5 - 5x + 3x =$

11) $(x - 3y)2 + 4y =$

12) $2.5x^2 \times (-5x) =$

13) $-4 - 2x^2 + 6x^2 =$

14) $8 + 14x^2 + 4 =$

15) $4(-2x - 7) + 10 =$

16) $(-x)(-2 + 3x) - x(7 + x) =$

17) $-3(6 + 12) - 3x + 5x =$

18) $-4(5 - 12x - 3x) =$

19) $3(-2x - 6) =$

20) $9 + 7x - 9 =$

21) $x(-2x + 8) =$

22) $5xy + 4x - 3y + x + 2y =$

23) $3(-x - 7) + 9 =$

24) $(-3x - 4) + 7 =$

25) $3x + 4y - 5 + 1 =$

26) $(-2 + 3x) - 3x(1 + 2x) =$

27) $(-3)(-3x - 3y) =$

28) $4(-x - 2) + 5 =$

Factoring Expressions

Factor the common factor out of each expression.

1) $12x - 6 =$

2) $5x - 15 =$

3) $\frac{45}{15}x - 15 =$

4) $7b - 28 =$

5) $4a^2 - 24a =$

6) $2xy - 10y =$

7) $5x^2y + 15x =$

8) $a^2 - 8a + 7ab =$

9) $2a^2 + 2ab =$

10) $4x + 20 =$

11) $24x - 36xy =$

12) $8x - 6 =$

13) $\frac{1}{4}x - \frac{3}{4}y =$

14) $7xy - \frac{14}{3}x =$

15) $3ab + 9c =$

16) $\frac{1}{3}x - \frac{4}{3} =$

17) $10x - 15xy =$

18) $x^2 + 8x =$

19) $4x^2 - 12y =$

20) $4x^3 + 3xy + x^2 =$

21) $21x - 14 =$

22) $20b - 60c + 20d =$

23) $24ab - 8ac =$

24) $ax - ay - 3x + 3y =$

25) $3ax + 4a + 9x + 12 =$

26) $x^2 - 10x =$

27) $9x^3 - 18x^2 =$

28) $5x^2 - 70xy =$

Evaluate One Variable Expressions

Evaluate each using the values given.

1) $x + 4x, x = 3$

2) $5(-6 + 3x), x = 1$

3) $4x + 7x, x = -3$

4) $5(2 - x) + 5, x = 3$

5) $6x + 4x - 10, x = 2$

6) $5x + 11x + 12, x = -1$

7) $5x - 2x - 4, x = 5$

8) $\frac{3(5x+8)}{9}, x = 2$

9) $2x - 85, x = 32$

10) $\frac{x}{18}, x = 108$

11) $7(3 + 2x) - 33, x = 5$

12) $7(x + 3) - 23, x = 4$

13) $\frac{x+(-6)}{-3}, x = -6$

14) $8(6 - 3x) + 5, x = 2$

15) $-11 - \frac{x}{5} + 3x, x = 10$

16) $5x + 11x, x = 1$

17) $-12x + 3(5 + 3x), x = -7$

18) $x + 11x, x = 0.5$

19) $\frac{(2x-2)}{6}, x = 13$

20) $3(-1 - 2x), x = 5$

21) $5x - (5 - x), x = 3$

22) $\left(-\frac{15}{x}\right) + 2 + x, x = 5$

23) $-\frac{x \times 5}{x}, x = 5$

24) $2(-1 - 3x), x = 2$

25) $2x^2 + 7x, x = 1$

26) $2(3x + 1) - 4(x - 5), x = 3$

27) $-6x - 4, x = -5$

28) $7x + 2x, x = 3$

Evaluate Two Variable Expressions

Evaluate the expressions.

1) $x + 4y$, $x = 5, y = 2$

2) $(-2)(-3x - 2y)$, $x = 1, y = 2$

3) $4x + 2y$, $x = 10, y = 5$

4) $\frac{x-4}{y+1}$, $x = 8, y = 3$

5) $\frac{a}{4} - 6b$, $a = 32, b = 4$

6) $3x - 4(y - 8)$, $x = 5, y = 3$

7) $3x + 2y - 10$, $x = 2, y = 10$

8) $-3x + 10 + 8y - 5$, $x = 2, y = 1$

9) $yx \div 3$, $x = 9, y = 9$

10) $a - b \div 3$, $a = 3, b = 12$

11) $6(x - y)$, $x = 7, y = 4$

12) $5x - 4y$, $x = 5, y = 8$

13) $\frac{10}{a} + 3b$, $a = 5, b = 4$

14) $2x^2 + 4xy$, $x = 3, y = 5$

15) $8 - \frac{xy}{10} + y$, $x = 6, y = 5$

16) $7(3x - y)$, $x = 7, y = -9$

17) $5x^2 - 3y^2$, $x = -1, y = 2$

18) $3x + \frac{y}{4}$, $x = 6, y = 16$

19) $4(4x - 2y)$, $x = 3, y = 5$

20) $4x(y - \frac{1}{2})$, $x = 5, y = 4$

21) $5(x^2 - 2y)$, $x = 3, y = 2$

22) $5xy$, $x = 2, y = 8$

23) $\frac{1}{3}y^3\left(y - \frac{1}{4}x\right)$, $x = -4, y = 3$

24) $-3(x - 5y) - 2x$, $x = 4, y = 2$

25) $-2x + \frac{1}{6}xy$, $x = 3, y = 6$

26) $x^2 + xy^2$, $x = 5, y = 7$

27) $x - 2y + 8$, $x = 9, y = 6$

28) $\frac{xy}{2x+y}$, $x = 5, y = 4$

Graphing Linear Equation

Sketch the graph of each line.

1) $y = 2x - 5$
2) $y = -2x + 3$
3) $x - y = 0$

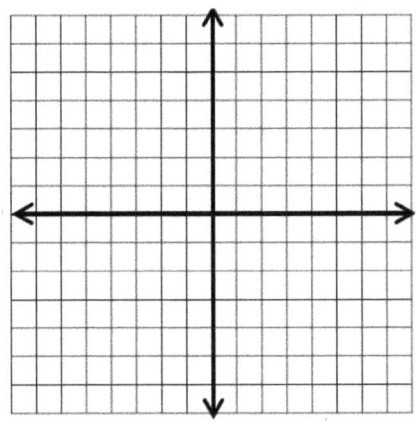
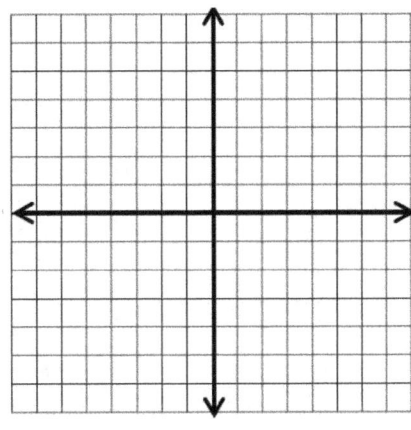
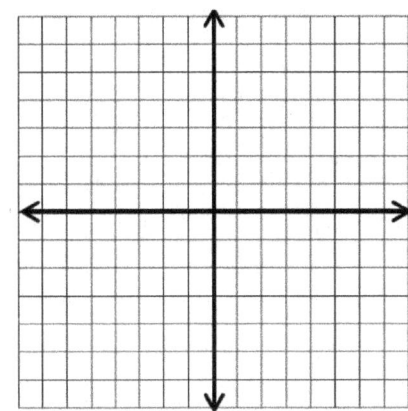

4) $x + y = 3$
5) $5x + 3y = -2$
6) $y - 3x + 2 = 0$

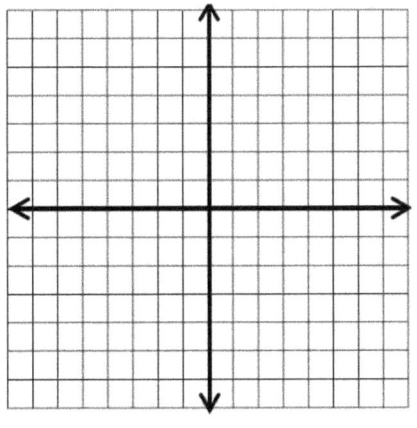
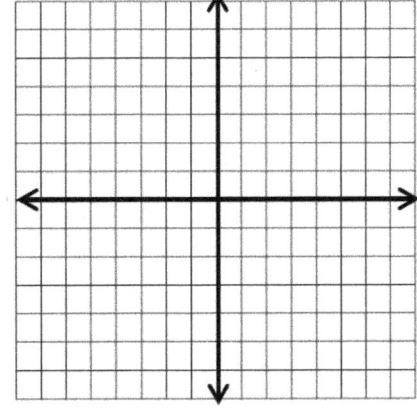
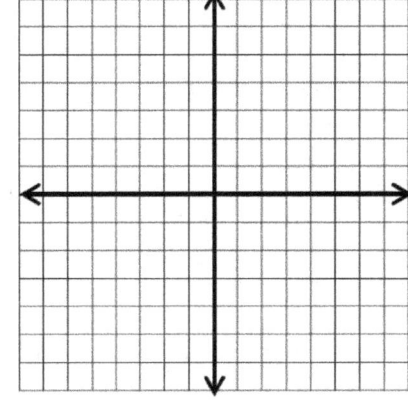

One Step Equations

Solve each equation.

1) $44 = (-12) + x$

2) $8x = (-64)$

3) $(-72) = (-8x)$

4) $(-5) = 3 + x$

5) $4 + \frac{x}{2} = (-3)$

6) $8x = (-104)$

7) $62 = x - 13$

8) $\frac{x}{3} = (-15)$

9) $x + 112 = 154$

10) $x - \frac{1}{3} = \frac{2}{3}$

11) $(-24) = x - 32$

12) $(-3x) = 39$

13) $(-169) = (13x)$

14) $-4x + 42 = 50$

15) $5x + 3 = 38$

16) $80 = (-8x)$

17) $3x + 7 = 19$

18) $24x = 144$

19) $x - 18 = 15$

20) $0.9x = 4.5$

21) $4x = 84$

22) $2x + 2.98 = 66.98$

23) $x + 9 = 6$

24) $x + 14 = 6$

25) $9x + 41 = 5$

26) $\frac{1}{4}x + 30 = 12$

Two Steps Equations

Solve each equation.

1) $6(3 + x) = 42$

2) $(-7)(x - 2) = 56$

3) $(-8)(3x - 4) = (-16)$

4) $5(2 + x) = -15$

5) $19(3x + 11) = 38$

6) $4(2x + 2) = 24$

7) $5(8 + 3x) = (-20)$

8) $(-5)(5x - 3) = 40$

9) $2x + 12 = 16$

10) $\frac{4x - 5}{5} = 3$

11) $(-3) = \frac{x + 4}{7}$

12) $80 = (-8)(x - 3)$

13) $\frac{x}{3} + 7 = 19$

14) $\frac{1}{4} = \frac{1}{2} + \frac{x}{4}$

15) $\frac{11 + x}{5} = (-6)$

16) $(-3)(10 + 5x) = (-15)$

17) $(-3x) + 12 = 24$

18) $\frac{x + 5}{5} = -5$

19) $\frac{x + 23}{8} = 3$

20) $(-4) + \frac{x}{2} = (-14)$

21) $-5 = \frac{x + 7}{8}$

22) $\frac{9x - 3}{6} = 4$

23) $\frac{2x - 12}{8} = 6$

24) $40 = (-5)(x - 8)$

Multi Steps Equations

Solve each equation.

1) $2 - (4 - 5x) = 3$

2) $-15 = -(4x + 7)$

3) $6x - 18 = (-2x) + 6$

4) $-32 = (-5x) - 11x$

5) $3(2 + 3x) + 3x = -30$

6) $5x - 18 = 2 + 2x - 7 + 2x$

7) $12 - 6x = (-36) - 3x + 3x$

8) $16 - 4x - 4x = 8 - 4x$

9) $8 + 7x + x = (-12) + 3x$

10) $(-3x) - 3(-2 + 4x) = 366$

11) $20 = (-200x) - 5 + 5$

12) $61 = 5x - 23 + 7x$

13) $7(4 + 2x) = 140$

14) $-60 = (-7x) - 13x$

15) $2(4x + 5) = -2(x + 4) - 22$

16) $11x - 17 = 6x + 8$

17) $9 = -3(x - 8)$

18) $(-6) - 8x = 6(1 + 2x)$

19) $x + 3 = -2(9 + 3x)$

20) $10 = 4 - 5x - 9$

21) $-15 - 9x - 3x = 12 - 3x$

22) $-23 - 3x + 5x = 27 - 23x$

23) $19 - 6x - 9x = -5 - 9x$

24) $15x - 18 = 6x + 9$

Graphing Linear Inequalities

Sketch the graph of each linear inequality.

1) $y > 2x - 3$
2) $y < x + 3$
3) $y \leq -3x - 8$

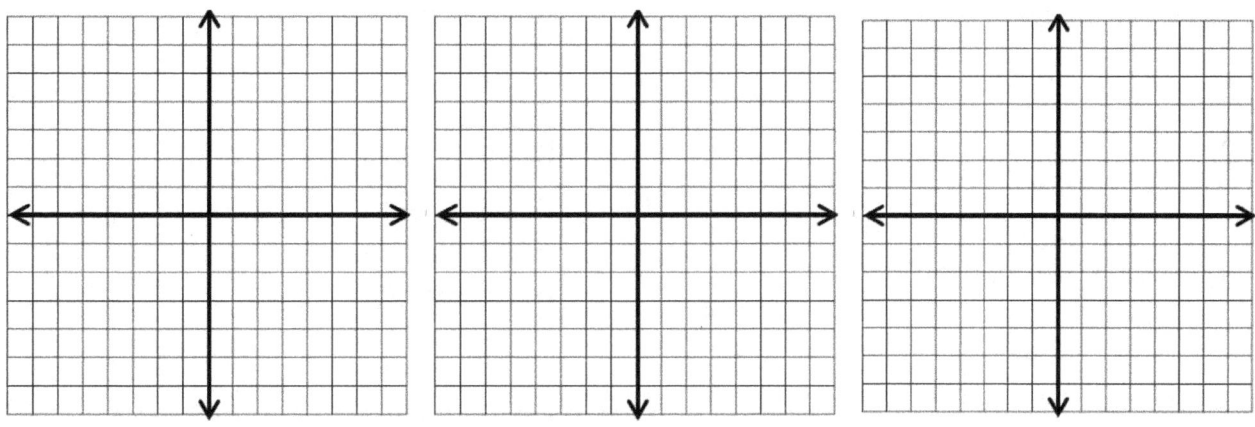

4) $3y \geq 6 + 3x$
5) $-3y < x - 12$
6) $2y \geq -8x + 4$

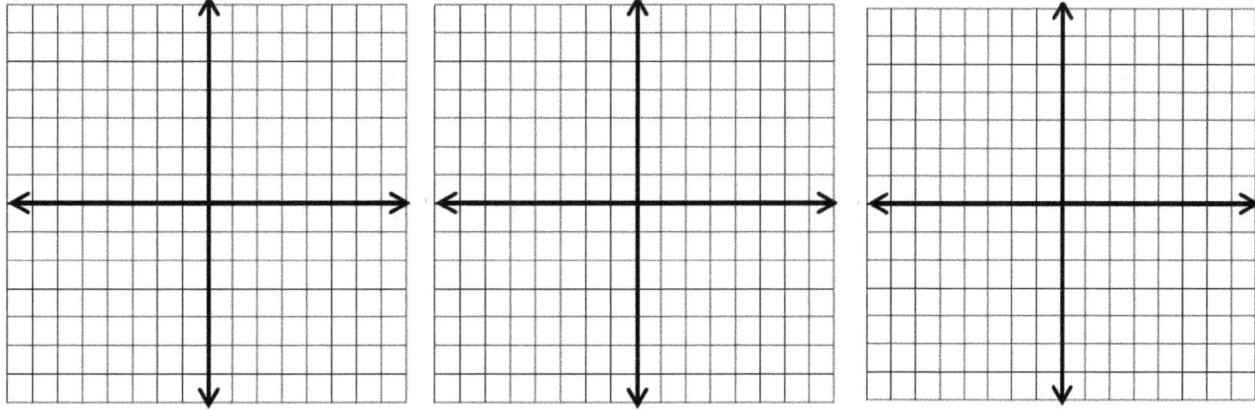

One Step Inequality

Solve each inequality.

1) $7x < 14$

2) $x + 7 \geq -8$

3) $x - 1 \leq 9$

4) $-2x + 4 > -10$

5) $x + 18 \geq -6$

6) $x + 9 \geq 5$

7) $x - \frac{1}{3} \leq 5$

8) $-7x < 42$

9) $-x + 8 > -3$

10) $\frac{x}{3} + 3 > -9$

11) $-x + 8 > -4$

12) $x - 14 \leq 18$

13) $-x - 5 \leq -7$

14) $x + 26 \geq -13$

15) $x + \frac{1}{3} \geq -\frac{2}{3}$

16) $x + 6 \geq -14$

17) $x - 42 \leq -48$

18) $x - 5 \leq 4$

19) $-x + 5 > -6$

20) $x + 6 \geq -12$

21) $8x + 6 \leq 22$

22) $4x - 3 \geq 9$

23) $3x - 5 < 22$

24) $6x - 8 \leq 40$

Two Steps Inequality

Solve each inequality

1) $2x - 3 \leq 7$

2) $3x - 4 \leq 8$

3) $\frac{-1}{4}x + \frac{x}{2} \leq \frac{1}{8}$

4) $5x + 10 \geq 30$

5) $4x - 7 \geq 9$

6) $3x - 5 \leq 16$

7) $8x - 2 \leq 14$

8) $9x + 5 \leq 23$

9) $2x + 10 > 32$

10) $\frac{x}{8} + 2 \leq 4$

11) $3x + 4 \geq 37$

12) $3x - 8 < 10$

13) $6 \geq \frac{x+7}{2}$

14) $3x + 9 < 48$

15) $\frac{4+x}{5} \geq 3$

16) $16 + 4x < 36$

17) $16 > 6x - 8$

18) $5 + \frac{x}{3} < 6$

19) $-4 + 4x > 24$

20) $5 + \frac{x}{7} < 3$

Multi Steps Inequality

Solve the inequalities.

1) $4x - 6 < 5x - 9$

2) $\frac{4x + 5}{3} \leq x$

3) $7x - 5 > 3x + 15$

4) $-3x > -6x + 4$

5) $3 + \frac{x}{2} < \frac{x}{4}$

6) $\frac{4x - 6}{8} > x$

7) $4x - 20 + 4 > 6x - 8$

8) $x - 8 > 11 + 3(x + 5)$

9) $\frac{x}{3} + 2 > x$

10) $-7x + 8 \geq -6(4x - 8) - 8x$

11) $7x - 4 \leq 8x + 9$

12) $\frac{2x - 7}{5} > 2$

13) $8(x + 2) < 6x + 10$

14) $-8x + 12 \leq 4(x - 9)$

15) $\frac{5x - 6}{3} > 3x + 2$

16) $2(x - 8) + 10 \geq 4x - 2$

17) $\frac{-5x+7}{6} > 5x$

18) $-3x - 4 > -7x$

19) $\frac{1}{4}x - 12 > \frac{1}{8}x - 19$

20) $-4(x - 9) \leq 5x$

Finding Distance of Two Points

Find the distance between each pair of points.

1) $(2, 1), (-1, -3)$
2) $(-4, -2), (4, 4)$
3) $(-3, 0), (15, 24)$
4) $(-4, -1), (1, 11)$
5) $(3, -2), (-6, -14)$
6) $(-6, 0), (-2, 3)$
7) $(3, 2), (11, 17)$
8) $(-6, -10), (6, -1)$
9) $(5, 9), (-11, -3)$
10) $(6, -2), (2, -6)$
11) $(3, 0), (18, 36)$
12) $(8, 4), (3, -8)$
13) $(4, 2), (-5, -10)$
14) $(-8, 10), (4, 40)$
15) $(8, 4), (-10, -20)$
16) $(-8, -2), (16, 8)$
17) $(3, 5), (-5, -10)$
18) $(-10, 20), (35, 45)$

Find the midpoint of the line segment with the given endpoints.

1) $(-2, -2), (4, 2)$
2) $(10, 4), (-2, 4)$
3) $(12, -2), (4, 10)$
4) $(-6, -5), (2, 1)$
5) $(3, -2), (5, -2)$
6) $(-10, -4), (6, -2)$
7) $(4, 1), (-4, 9)$
8) $(-5, 6), (-5, 2)$
9) $(-8, 8), (4, -2)$
10) $(1, 7), (5, -1)$
11) $(-9, 5), (5, 3)$
12) $(7, 10), (-3, -6)$
13) $(-8, 14), (-8, 2)$
14) $(16, 7), (6, -3)$
15) $(5, 6), (-3, 4)$
16) $(-9, -1), (-5, 7)$
17) $(17, 9), (5, 11)$
18) $(-8, -11), (18, -1)$

Answer key Chapter 8

Distributive and Simplifying Expressions

1) $6x - 6$
2) $4 + 5x$
3) $6x - 8$
4) $-2x^2 - 6x$
5) $5x^2 - 2x$
6) $2x + 15y$
7) $11x - 6y$
8) $6x - 1$
9) $-8x + 21$
10) $-2x - 5$
11) $2x - 2y$
12) $-12.5x^3$
13) $4x^2 - 4$
14) $14x^2 + 12$
15) $-8x - 18$
16) $-4x^2 - 5x$
17) $2x - 54$
18) $60x - 20$
19) $-6x - 18$
20) $7x$
21) $-2x^2 + 8x$
22) $5x + y + 5xy$
23) $-3x - 12$
24) $-3x + 3$
25) $3x + 4y - 4$
26) $-6x^2 - 2$
27) $9x + 9y$
28) $-4x - 3$

Factoring Expressions

1) $3(4x - 2)$
2) $5(x - 3)$
3) $3(x - 5)$
4) $7(b - 4)$
5) $4a(a - 6)$
6) $2y(x - 5)$
7) $5x(xy + 3)$
8) $a(a - 8 + 7b)$
9) $2a(a + b)$
10) $4(x + 5)$
11) $12x(2 - 3y)$
12) $2(4x - 3)$
13) $\frac{1}{4}(x - 3y)$
14) $7x(y - \frac{2}{3})$
15) $3(ab + 3c)$
16) $\frac{1}{3}(x - 4)$
17) $5x(2 - 3y)$
18) $x(x + 8)$
19) $4(x^2 - 3y)$
20) $x(4x^2 + 3y + x)$
21) $7(3x - 2)$
22) $20(b - 3c + d)$
23) $8a(3b - c)$
24) $(x - y)(a - 3)$
25) $(3x + 4)(a + 3)$
26) $x(x - 10)$
27) $9x^2(x - 2)$
28) $5x(x - 14y)$

Evaluate One Variable Expressions

1) 15
2) −15
3) −33
4) 0
5) 10
6) −4
7) 11
8) 6
9) −21
10) 6
11) 58
12) 26
13) 4
14) 5
15) 17
16) 16
17) 36
18) 6
19) 4
20) −33
21) 13
22) 4
23) −5
24) −14

WWW.MathNotion.com

SBAC Math Practice Grade 7

25) 9   26) 28   27) 26   28) 27

Evaluate Two Variable Expressions

1) 13
2) 14
3) 50
4) 1
5) −16
6) 35
7) 16
8) 7
9) 27
10) 3
11) 18
12) 20
13) 17
14) 78
15) 10
16) 210
17) −7
18) 22
19) 8
20) 70
21) 25
22) 80
23) 36
24) 10
25) −3
26) 270
27) 5
28) $\frac{10}{7}$

Graphing Lines Using Line Equation

1) $y = 2x - 5$

2) $y = -2x + 3$

3) $x - y = 0$

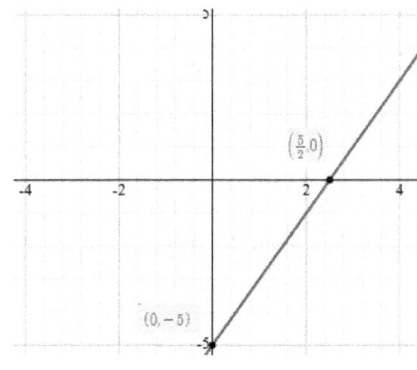
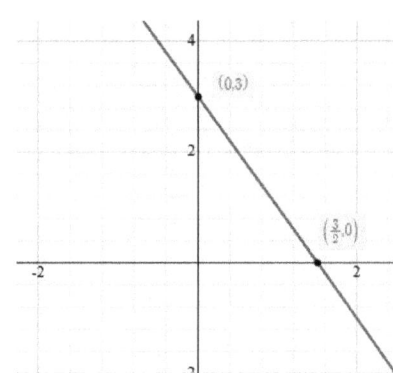
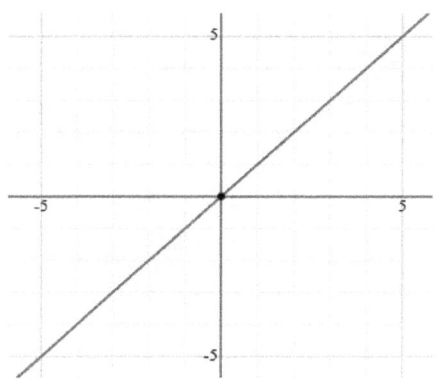

4) $x + y = 3$

5) $5x + 3y = -2$

6) $y - 3x + 2 = 0$

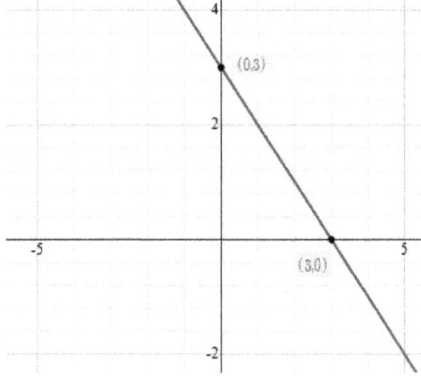
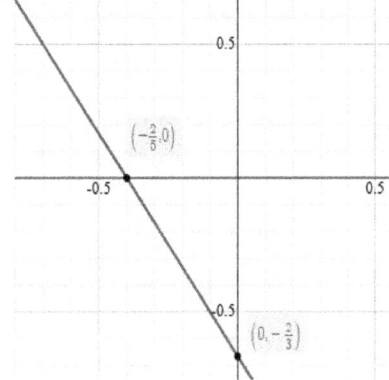
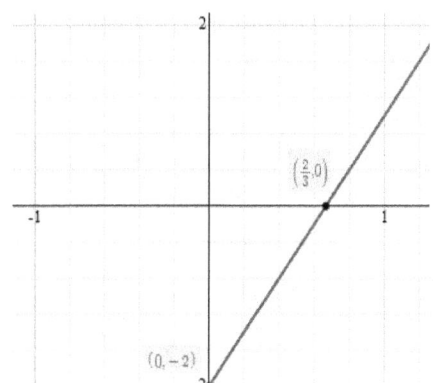

SBAC Math Practice Grade 7

One Step Equations

1) $x = 56$
2) $x = -8$
3) $x = 9$
4) $x = -8$
5) $x = -14$
6) $x = -13$
7) $x = 75$
8) $x = -45$
9) $x = 42$
10) $x = 1$
11) $x = 8$
12) $x = -13$
13) $x = -13$
14) $x = -2$
15) $x = 7$
16) $x = -10$
17) $x = 4$
18) $x = 6$
19) $x = 33$
20) $x = 5$
21) $x = 21$
22) $x = 32$
23) $x = -3$
24)
25)
26)

Two Steps Equations

1) $x = 4$
2) $x = -6$
3) $x = 2$
4) $x = -5$
5) $x = -3$
6) $x = 2$
7) $x = -4$
8) $x = -1$
9) $x = 2$
10) $x = 5$
11) $x = -25$
12) $x = -7$
13) $x = 36$
14) $x = -1$
15) $x = -41$
16) $x = -1$
17) $x = -4$
18) $x = -30$
19) $x = 1$
20) $x = -20$
21) $x = -47$
22) $x = 3$
23) $x = 30$
24) $x = 0$

Multi Steps Equations

1) $x = 1$
2) $x = 2$
3) $x = 3$
4) $x = 2$
5) $x = -3$
6) $x = 13$
7) $x = 8$
8) $x = 2$
9) $x = -4$
10) $x = -24$
11) $x = -0.1$
12) $x = 7$
13) $x = 8$
14) $x = 3$
15) $x = -4$
16) $x = 5$
17) $x = 5$
18) $x = -3/5$
19) $x = -3$
20) $x = -3$
21) $x = -3$
22) $x = 2$
23) $x = 4$
24) $x = 3$

WWW.MathNotion.com

SBAC Math Practice Grade 7

Graphing Linear Inequalities

1) $y > 2x - 3$ 2) $y < x + 3$ 3) $y \leq -3x - 8$

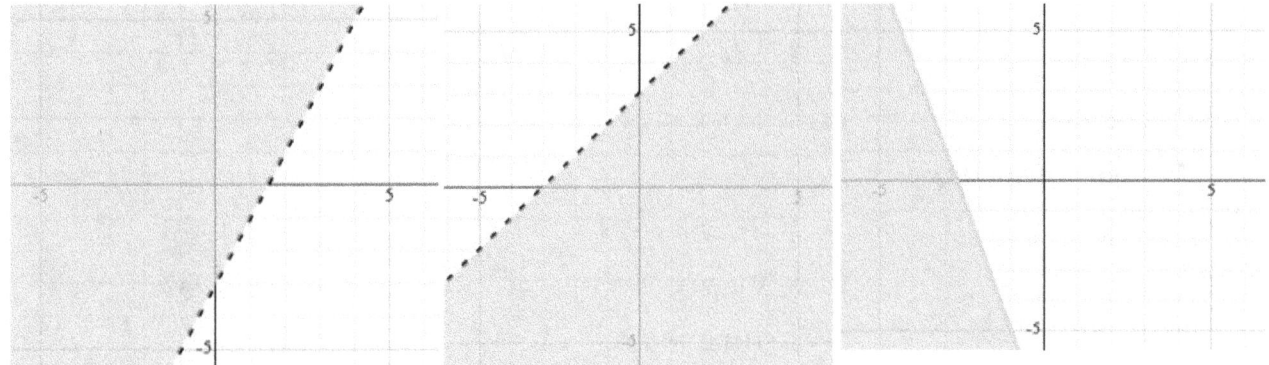

4) $3y \geq 6 + 3x$ 5) $-3y < x - 12$ 6) $2y \geq -8x + 4$

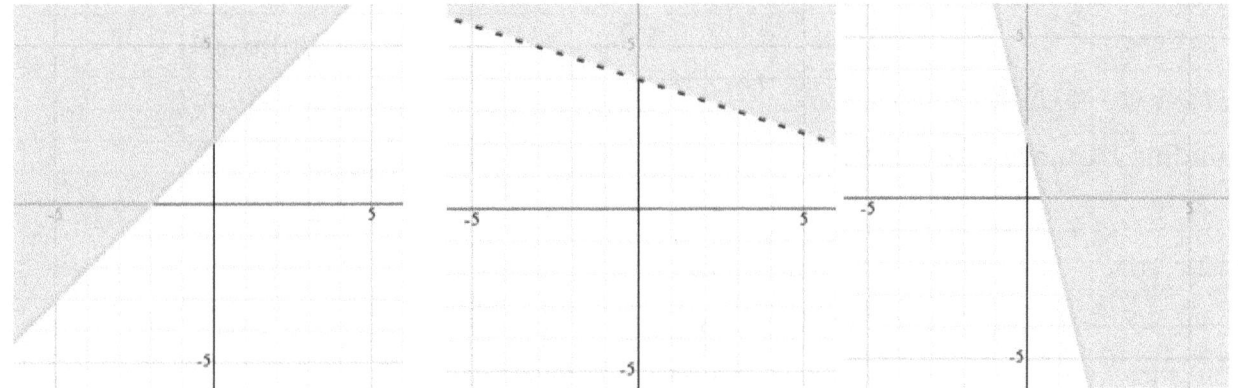

One Step Inequality

1) $x < 2$
2) $x \geq -15$
3) $x \leq 10$
4) $x < 7$
5) $x \geq -24$
6) $x \geq -4$
7) $x \leq \frac{16}{3}$
8) $x > -6$

9) $x < 11$
10) $x > -36$
11) $x < 12$
12) $x \leq 32$
13) $x \geq 2$
14) $x \geq -39$
15) $x \geq -1$
16) $x \geq -20$

17) $x \leq -6$
18) $x \leq 9$
19) $x < 11$
20) $x \geq -18$
21) $x \leq 2$
22) $x \geq 3$
23) $x < 9$
24) $x \leq 8$

Two Steps Inequality

1) $x \leq 5$
2) $x \leq 4$

3) $x \leq 0.5$
4) $x \geq 4$

5) $x \geq 4$
6) $x \leq 7$

WWW.MathNotion.com

7) $x \leq 2$
8) $x \leq 2$
9) $x > 11$
10) $x \leq 16$
11) $x \geq 11$

12) $x < 6$
13) $x \leq 5$
14) $x < 13$
15) $x \geq 11$
16) $x < 5$

17) $x < 4$
18) $x < 3$
19) $x > 7$
20) $x < -14$

Multi Steps Inequality

1) $x > 3$
2) $x \leq -5$
3) $x > 5$
4) $x > \frac{4}{3}$
5) $x < -12$
6) $x < -1.5$
7) $x < -4$

8) $x < -17$
9) $x < 3$
10) $x \geq 1.6$
11) $x \geq -13$
12) $x > 8.5$
13) $x < -3$
14) $x \geq 4$

15) $x < -3$
16) $x \leq -2$
17) $x < \frac{1}{5}$
18) $x > 1$
19) $x > -56$
20) $x \geq 4$

Finding Distance of Two Points

1) 5
2) 10
3) 30
4) 13
5) 15
6) 5

7) 17
8) 15
9) 20
10) $4\sqrt{2}$
11) 39
12) 13

13) 15
14) $6\sqrt{29}$
15) 30
16) 26
17) 17
18) $5\sqrt{106}$

Finding Midpoint

1) $(1, 0)$
2) $(4, 4)$
3) $(8, 4)$
4) $(-2, -2)$
5) $(4, -2)$
6) $(-2, -3)$

7) $(0, 5)$
8) $(-5, 4)$
9) $(-2, 3)$
10) $(3, 3)$
11) $(-2, 4)$
12) $(2, 2)$

13) $(-8, 8)$
14) $(11, 2)$
15) $(1, 5)$
16) $(-7, 3)$
17) $(11, 10)$
18) $(5, -6)$

Chapter 9: Transformations

Translations

Graph the image of the figure using the transformation given.

1) translation: 4 units right and 3 units down

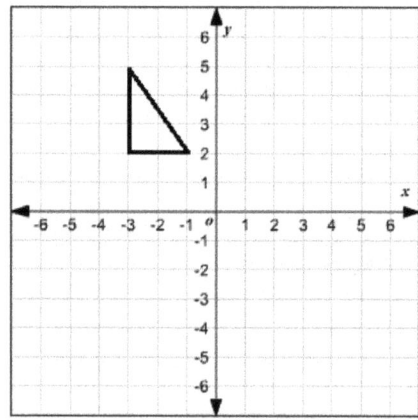

2) translation: 2 units left and 1 units down

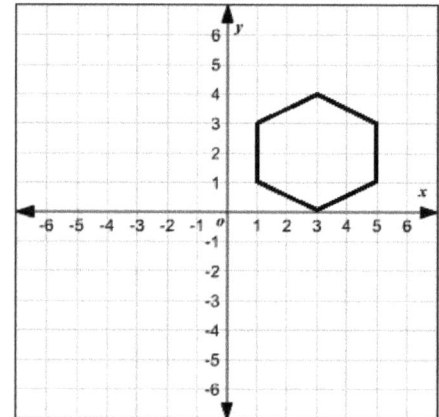

Write a rule to describe each transformation.

3)

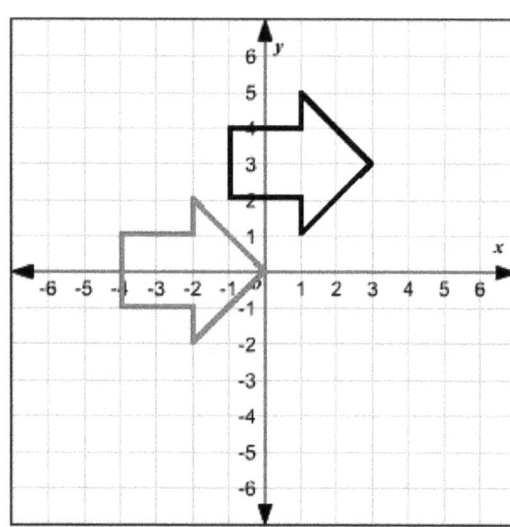

4)

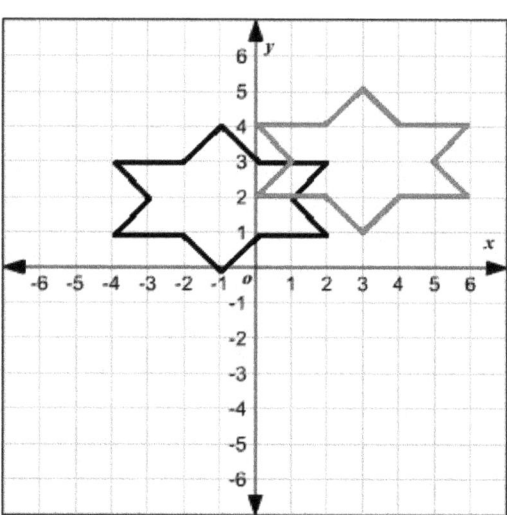

Reflections

Graph the image of the figure using the transformation given.

1) Reflection across $x = -2$

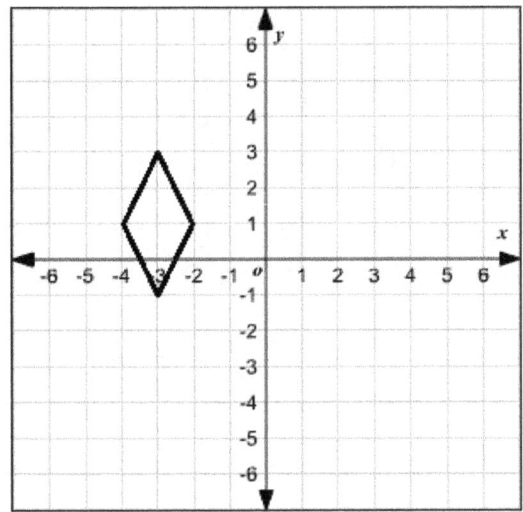

2) Reflection across $y = -x$

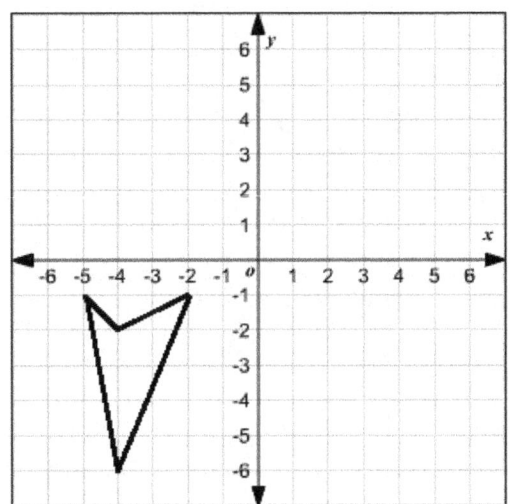

3) Reflection across $y = -1$

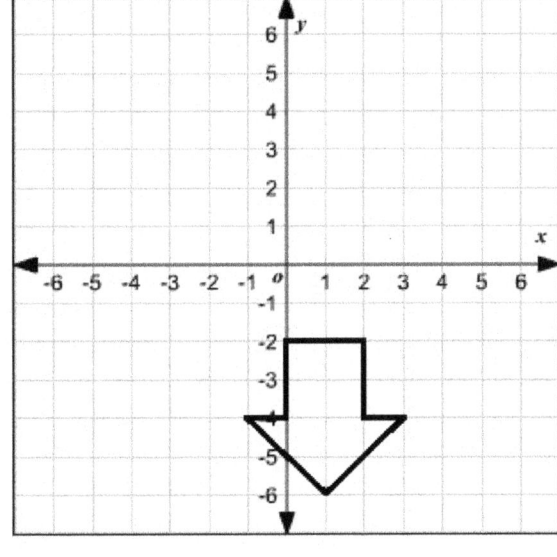

4) Reflection across x axis

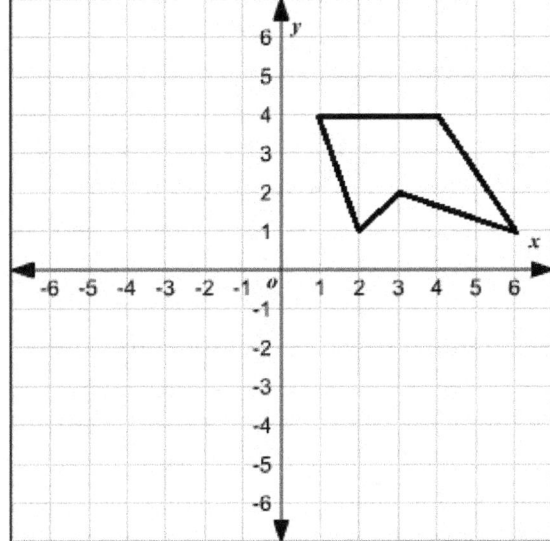

SBAC Math Practice Grade 7

Write a rule to describe each transformation.

5)
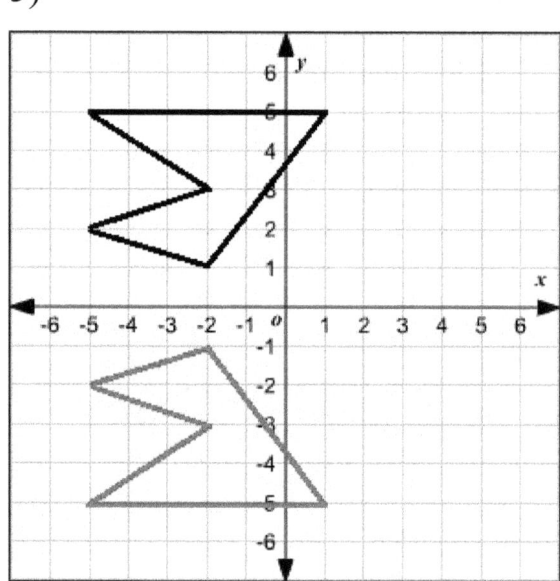

6)
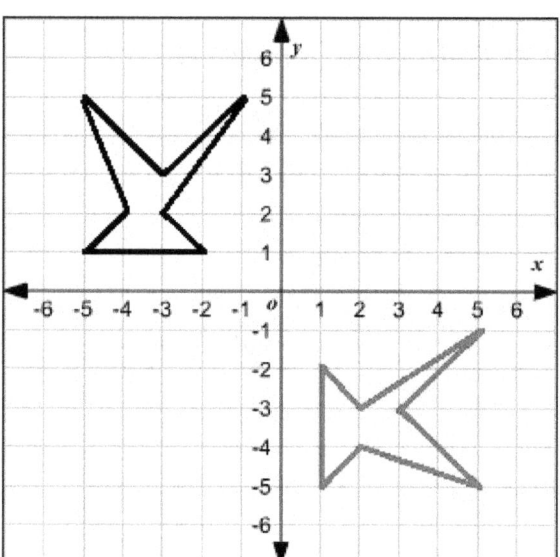

7)
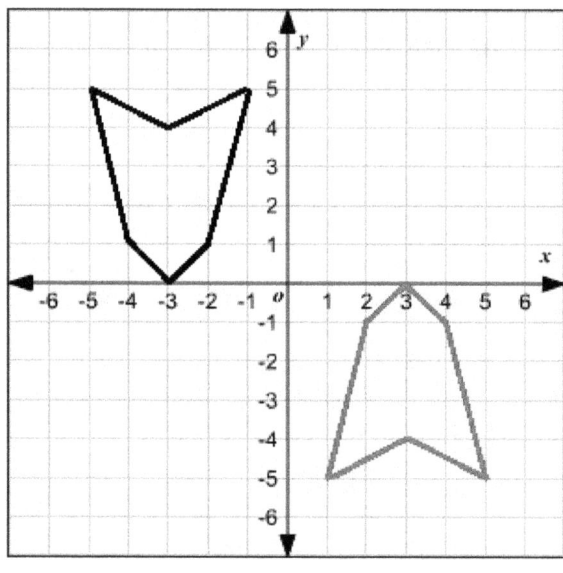

8)
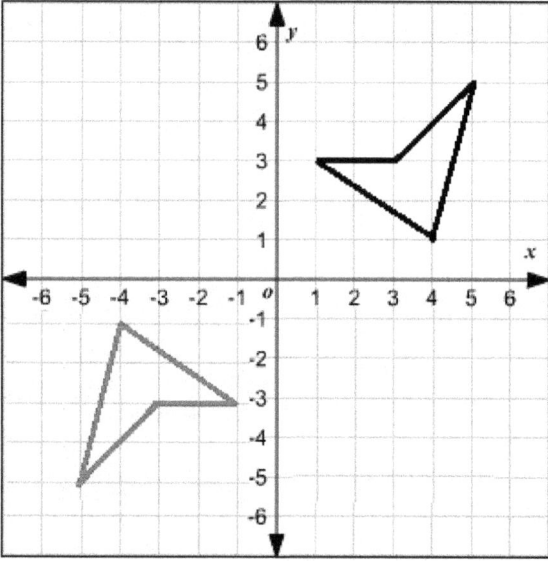

Rotations

Graph the image of the figure using the transformation given.

1) rotation 270° clockwise about the origin

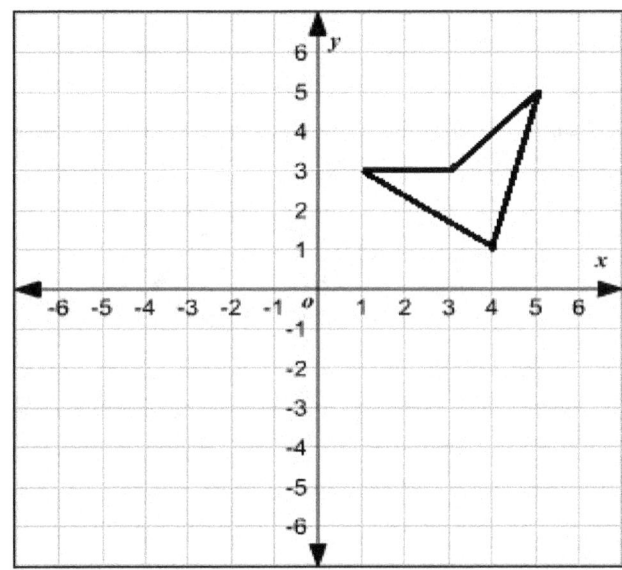

2) rotation 180° clockwise about the origin

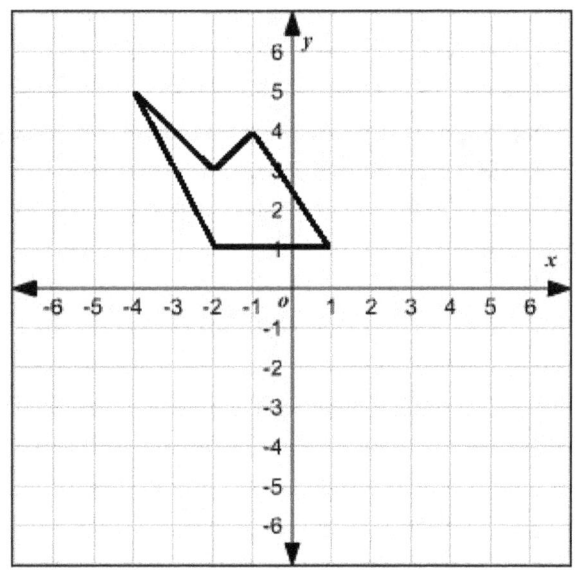

3) rotation 90° clockwise about the origin

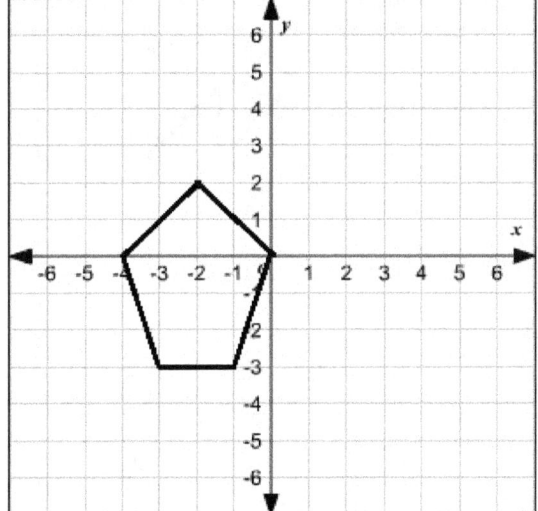

4) rotation 90° counterclockwise about the origin

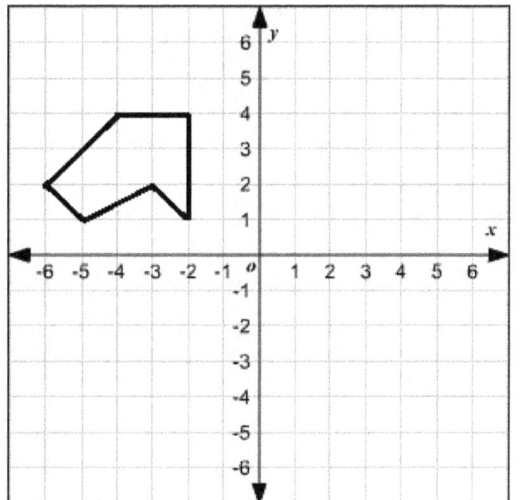

SBAC Math Practice Grade 7

Write a rule to describe each transformation.

5)

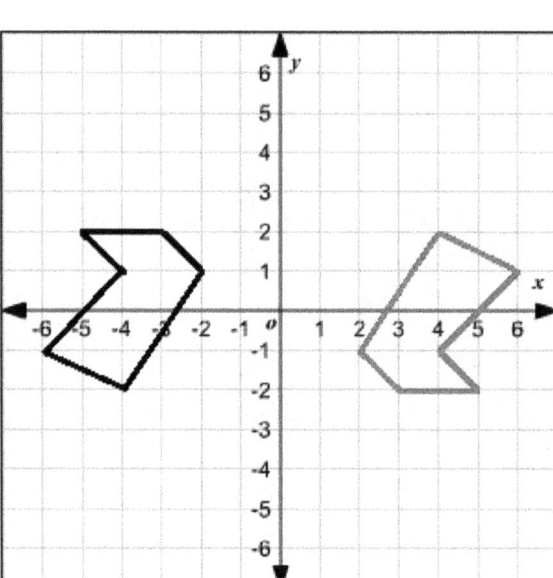

6)

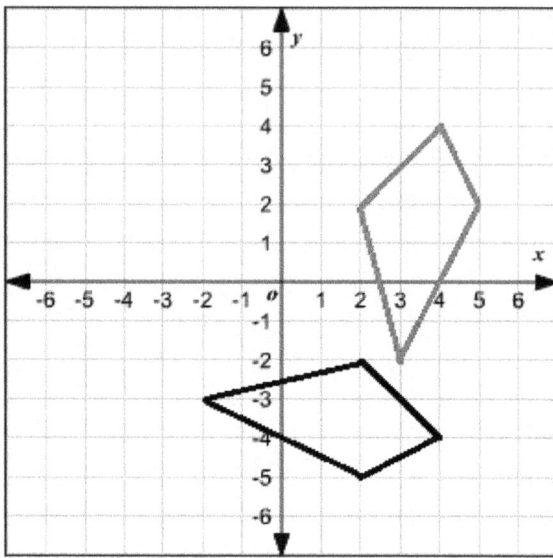

7)

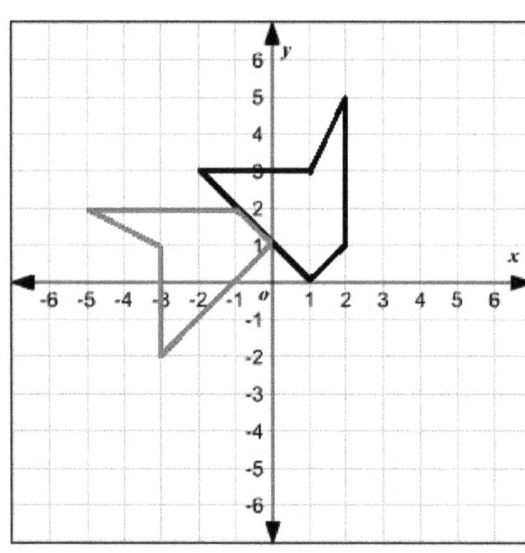

8)

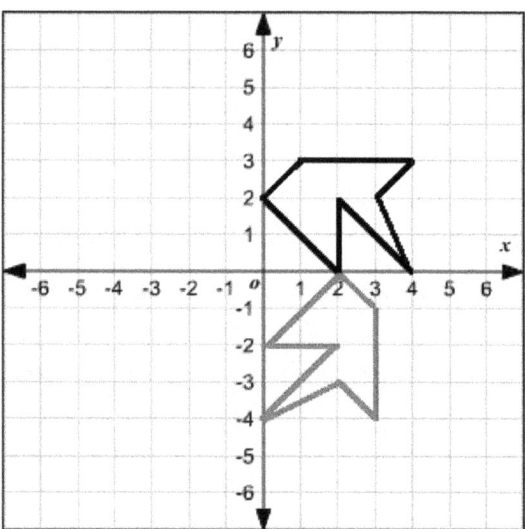

Dilations

Draw a dilation of the figure using the given scale factor.

1) $k = \frac{1}{3}$

2) $k = 2$

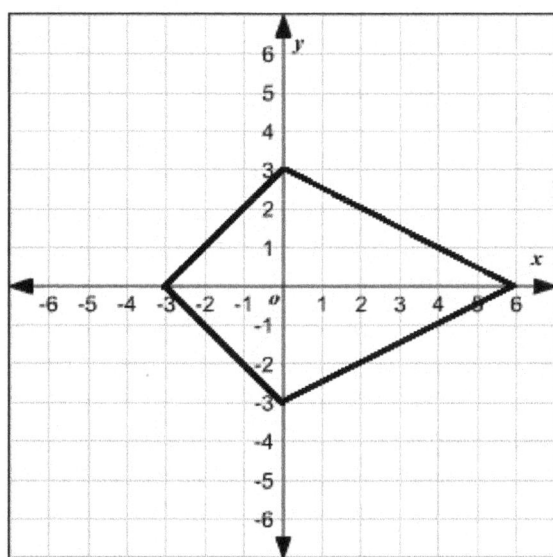
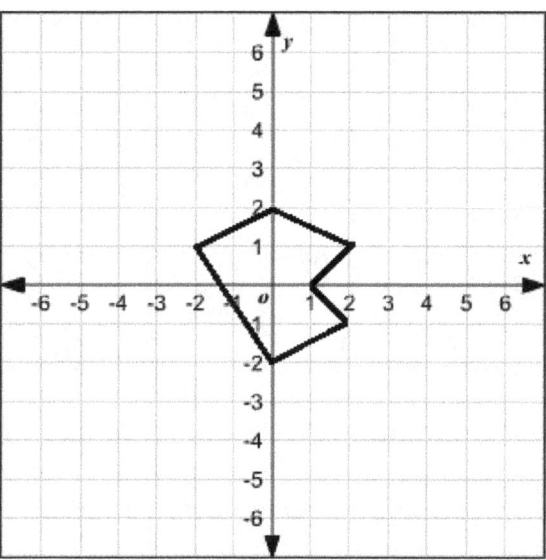

Determine whether the dilation from figure M to figure N is a reduction or an enlargement. Then find the scale factor and the missing length.

3)

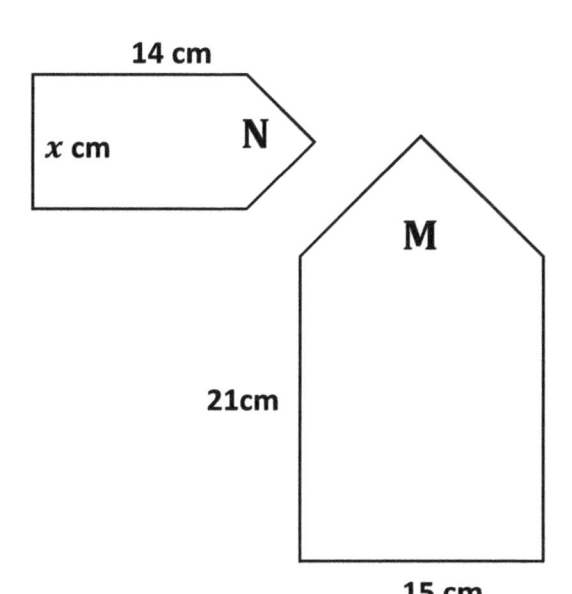

4)

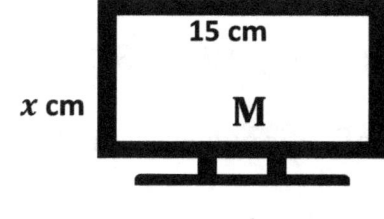

Coordinates of Vertices

Calculate the new coordinates after the given transformations.

1) Translate: 1-unit right and 3 units down.

   $A(-3, 1)$, $B(-1, -2)$, $C(3, 0)$

2) Rotation: 180° clockwise about the origin.

   $D(-2, 2)$, $E(-6, 10)$, $F(-8, 4)$, $G(-4, 12)$

3) Rotation: 90° counterclockwise about the origin.

   $P(1, 2)$, $Q(4, 1)$, $R(5, 5)$, $S(0, -2)$

4) Rotation: 270° counterclockwise about the origin.

   $J(-3, 4)$, $K(-6, 2)$, $L(1, -3)$

5) Reflection: over the x axis.

   $C(-1, -4)$, $D(-5, -2)$, $W(2, -6)$, $Y(8, 3)$

6) Reflection: across the line $y = x$.

   $A(4, -1)$, $B(6, -3)$, $C(5, -5)$, $D(3, -4)$

7) Reflection: across the line $y = -3$.

   $K(-2, -1)$, $L(-6, 0)$, $M(-3, -2)$, $N(-8, 1)$

8) Dilate: Reduction by scale factor $\frac{1}{4}$.

   $A(8, 2)$, $B(-10, -4)$, $C(-12, 8)$

9) Dilate: Enlargement by scale factor 3.

   $F(-2, 0)$, $G(-1, 2)$, $H(2, 2)$

Answers of Worksheets – Chapter 9

Translations

1)
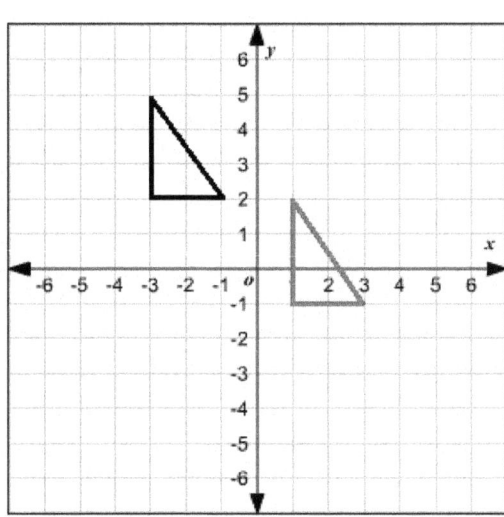

2)
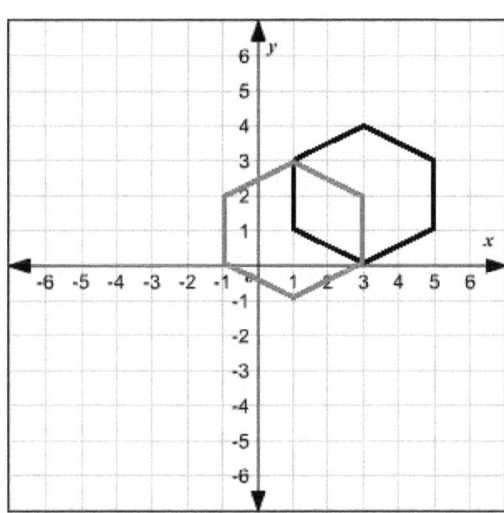

3) translation: : 3 units left and 3 units down

4) translation: 4 units right and 1 unit up

Reflections

1)
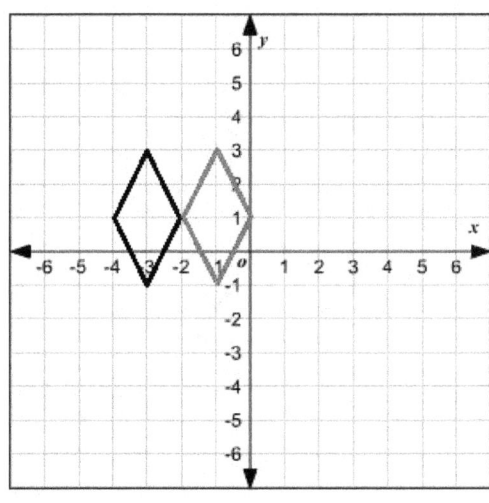

2)
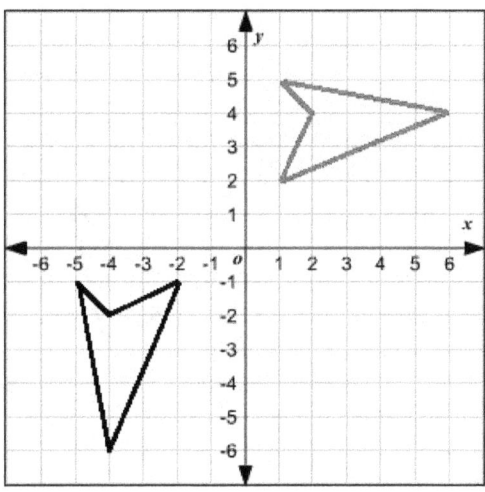

3)

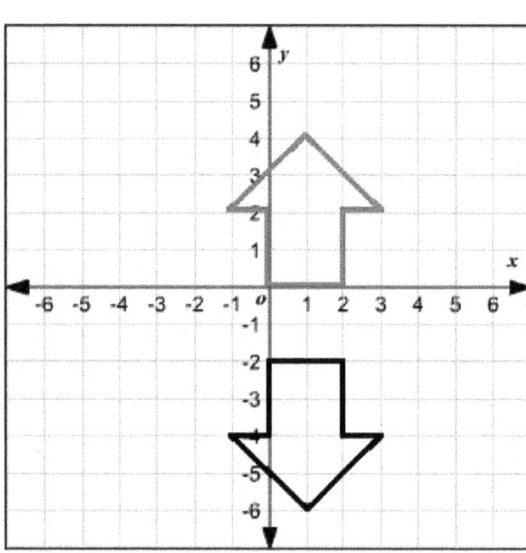

4)

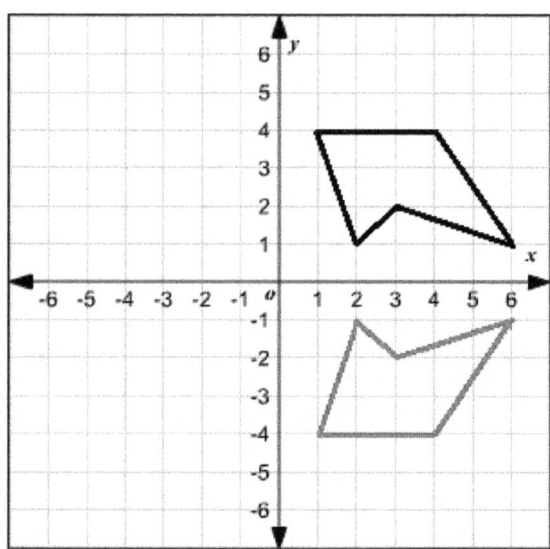

5) reflection across the y = 0 (x axis)

6) reflection across the y = x

7) reflection against the origin

8) reflection against the origin

Rotations

1)

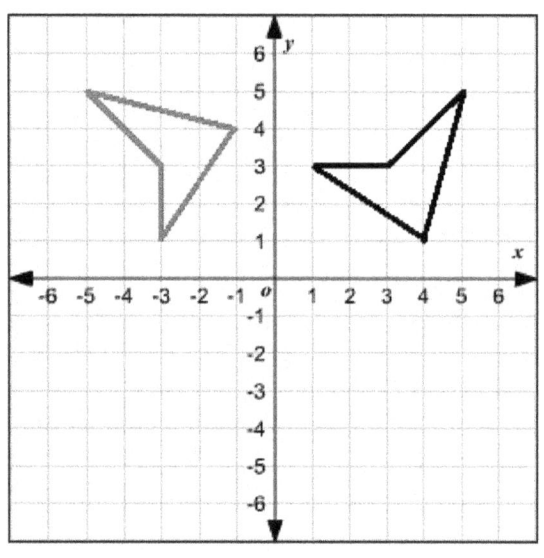

2)

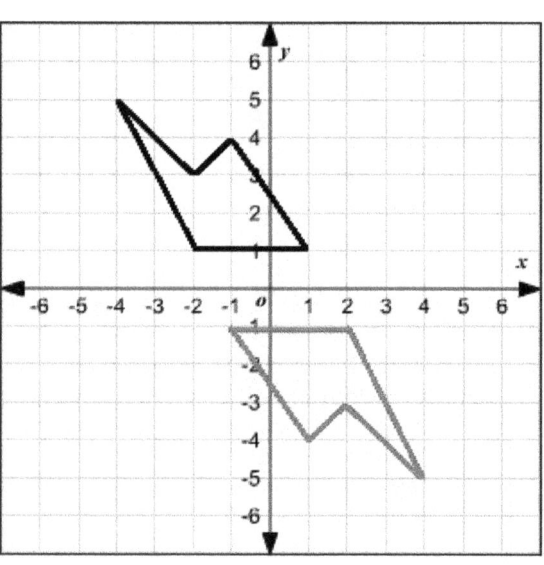

SBAC Math Practice Grade 7

3)

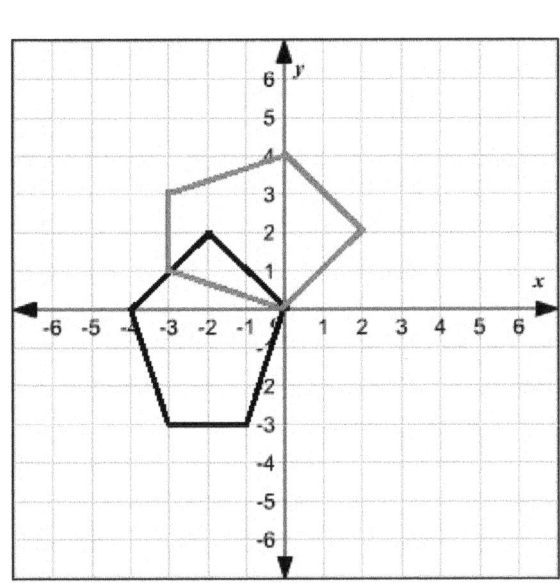

4)

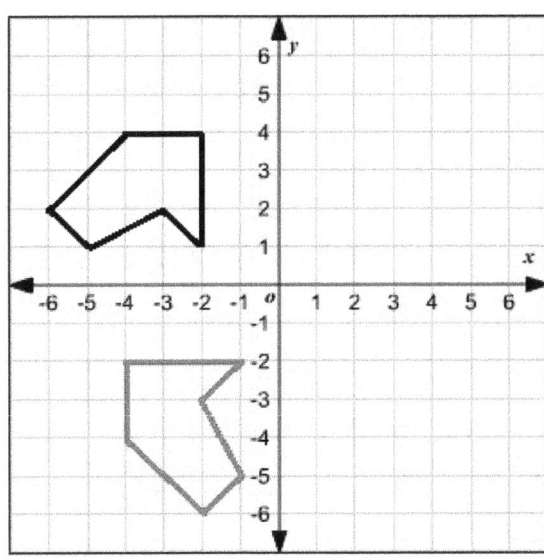

5) rotation 180° counter clockwise about the origin

6) rotation 270° about the origin

7) rotation 90° counter clockwise about the origin

8) rotation 270° counter clockwise about the origin

Dilations

1)

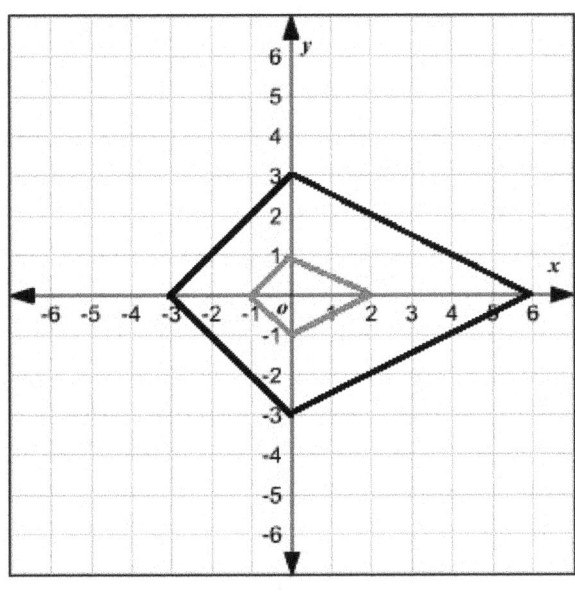

2)

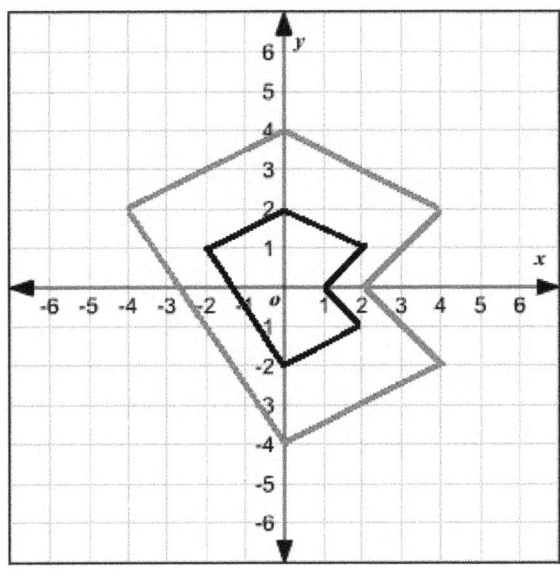

3) Reduction, $k = \frac{3}{2}$, $x = 10\ cm$

4) Enlargement, $k = \frac{1}{2}$, $x = 7.5\ cm$

Coordinate of Vertices

1) $A'(-2,-2), B'(0,-5), C'(4,-3)$
2) $D'(2,-2), E'(6,-10), F'(8,-4), G'(4,-12)$
3) $P'(-2,1), Q'(-1,4), R'(-5,5), S'(2,0)$
4) $J'(4,3), K'(2,6), L'(-3,-1)$
5) $C'(-1,4), D'(-5,2), W'(2,6), Y'(8,-3)$
6) $A'(-1,4), B'(-3,6), C'(-5,5), D'(-4,3)$
7) $K'(-2,-5), L'(-6,-6), M'(-3,-4), N'(-8,-7)$
8) $A'(2,0.5), B'(-2.5,-1), C'(-3,2)$
9) $F'(-6,0), G'(-3,6), H'(6,6)$

Chapter 10:
Geometry

Area and Perimeter of Square

Find the perimeter and area of each squares.

1) 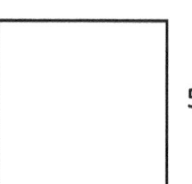 5

Perimeter: ....:

Area: ....:

2) 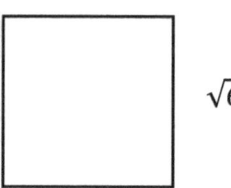 $\sqrt{6}$

Perimeter: ....:

Area: ....:

3) 7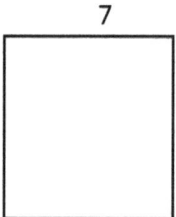

Perimeter: ....:

Area: ....:

4) $\sqrt{9}$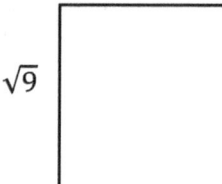

Perimeter: ....:

Area: ....:

5) 12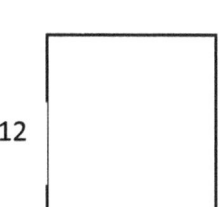

Perimeter: ....:

Area: ....:

6) 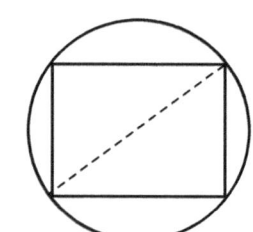 d=10

Perimeter of Square: ....:

Area of Square: ....:

Area and Perimeter of Rectangle

Find the perimeter and area of each rectangle.

1)

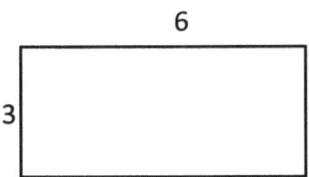

Perimeter: _____.

Area: _____.

2)

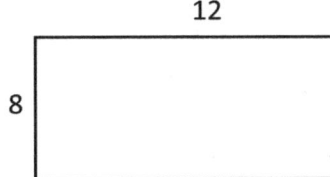

Perimeter: _____.

Area: _____.

3)

15
10

Perimeter: _____.

Area: _____.

4)

7
2.5

Perimeter: _____.

Area: _____.

5)

$1\frac{2}{5}$
$\frac{5}{7}$

Perimeter: _____.

Area: _____.

6)

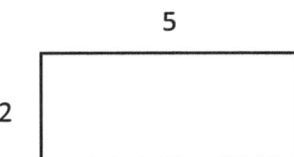

Perimeter: _____.

Area: _____.

Area and Perimeter of Triangle

Find the perimeter and area of each triangle.

1)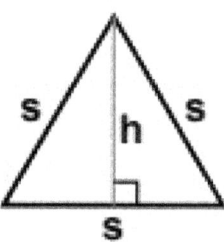

Perimeter:_____.

Area:_____:

2)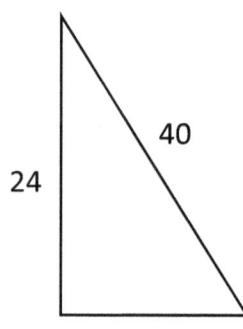

Perimeter:_____:

Area:_____:

3)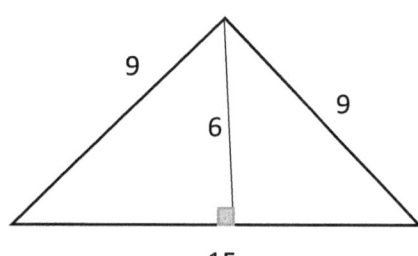

Perimeter:_____:

Area _____:

4) s=8

   h=6

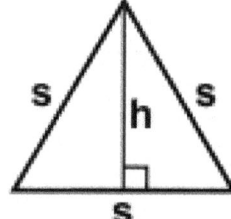

Perimeter:_____:

Area:_____.

5)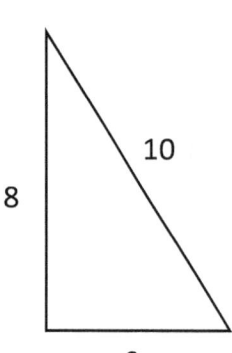

Perimeter:_____:

Area:_____:

6)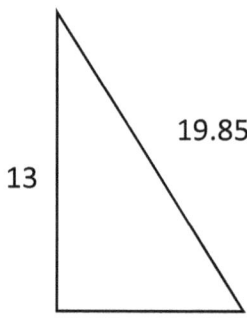

Perimeter:_____:

Area:_____:

Area and Perimeter of Trapezoid

Find the perimeter and area of each trapezoid.

1)
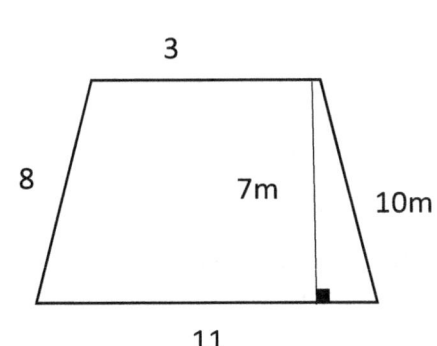

Perimeter: ....:

Area: ....:

2)
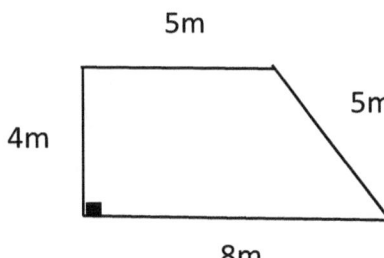

Perimeter: ....:

Area: ....:

3)
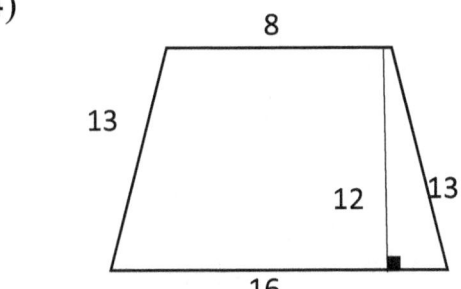

Perimeter: ....:

Area ....:

4)

Perimeter: ....:

Area: ....:

5)
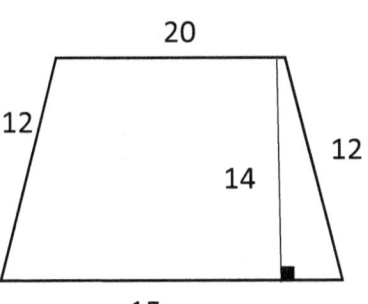

Perimeter: ....:

Area: ....:

6)
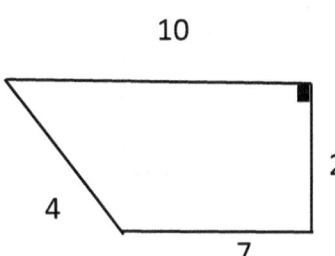

Perimeter: ....:

Area: ....:

Area and Perimeter of Parallelogram

Find the perimeter and area of each parallelogram.

1)

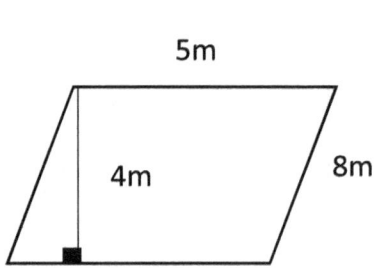

Perimeter: _____ :

Area: _____ :

2)

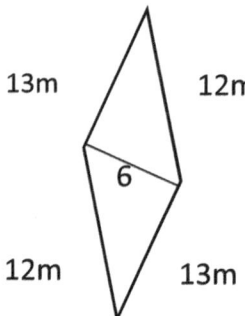

Perimeter: _____ :

Area: _____ :

3)

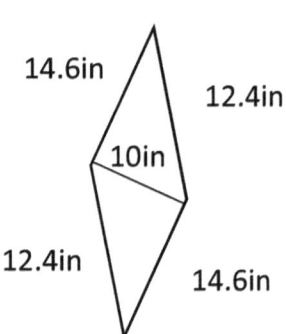

Perimeter: _____ :

Area _____ :

4)

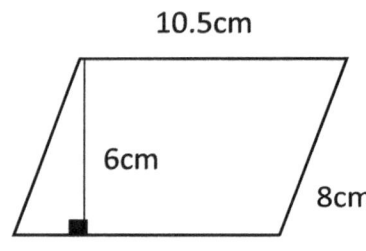

Perimeter: _____ :

Area: _____ :

5)

24.5m

18m

Perimeter: _____ :

Area: _____ :

6)

12 m

Perimeter: _____ :

Area: _____ :

Circumference and Area of Circle

Find the circumference and area of each ($\pi = 3.14$).

1)

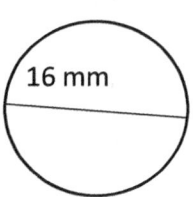

Circumference:

Area:

2)

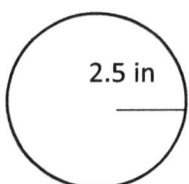

Circumference: _____

Area: _____

3)

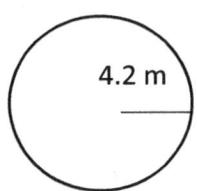

Circumference: _____

Area _____

4)

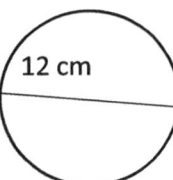

Circumference: _____

Area: _____

5)

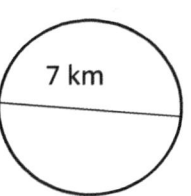

Circumference: _____

Area: _____

6)

Circumference: _____

Area: _____

WWW.MathNotion.com

Perimeter of Polygon

Find the perimeter of each polygon.

1)

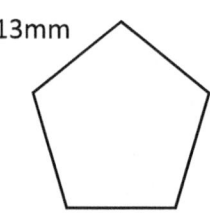

Perimeter: _____ .

2)

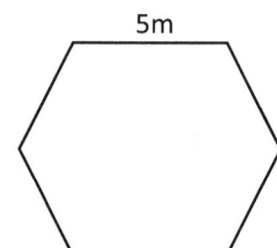

Perimeter: _____ :

3)

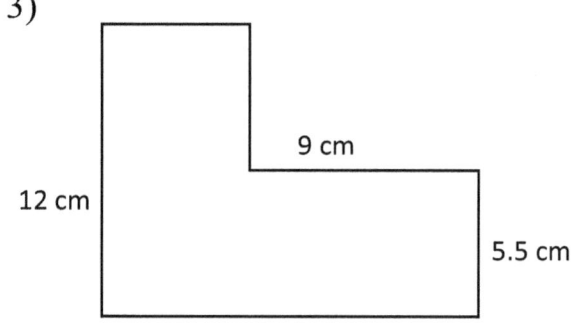

Perimeter: _____ .

4)

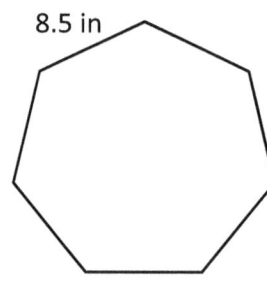

Perimeter: _____ :

5)

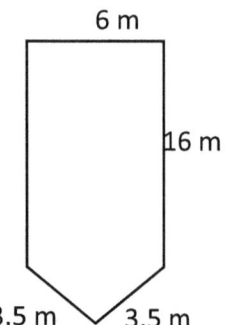

Perimeter: _____ .

6)

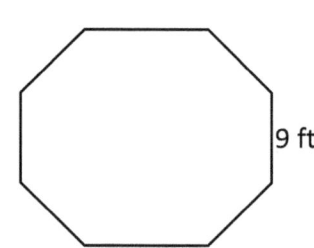

Perimeter: _____ :

Volume of Cubes

Find the volume of each cube.

1)
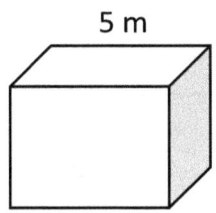

V: _____ .

2)
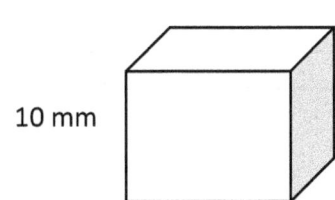

V: _____ .

3)
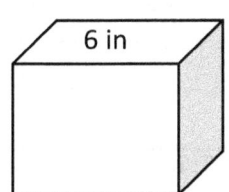

V: _____ .

4)
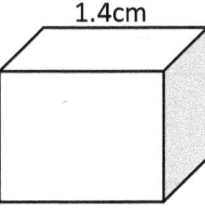

V: _____ .

5)
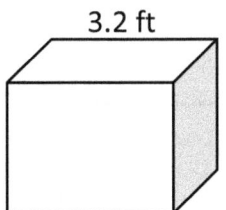

V: _____ .

6)
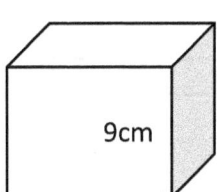

V: _____ .

Volume of Rectangle Prism

Find the volume of each rectangle prism

1)
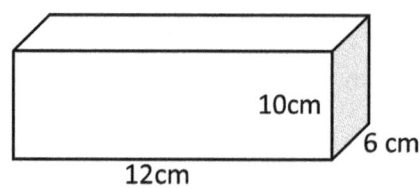

V:……………………….

2)
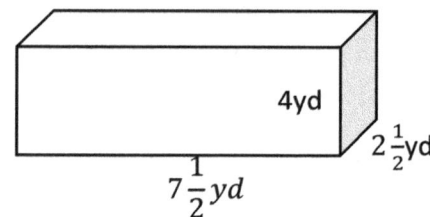

V:……………………….

3)
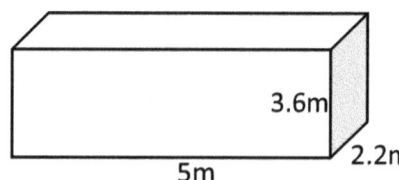

V:……………………….

4)
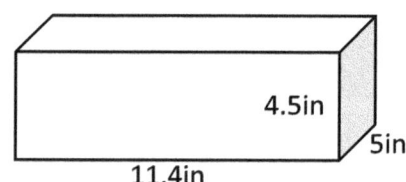

V:……………………….

5)
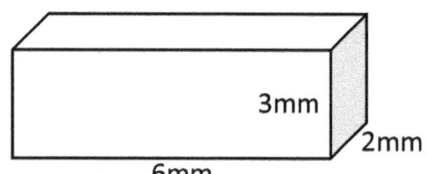

V:……………………….

6)
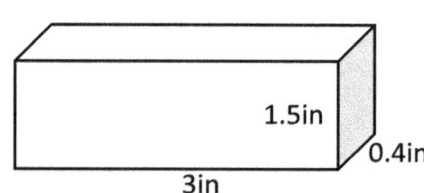

V:……………………….

Volume of Cylinder

Find the volume of each cylinder.

1)
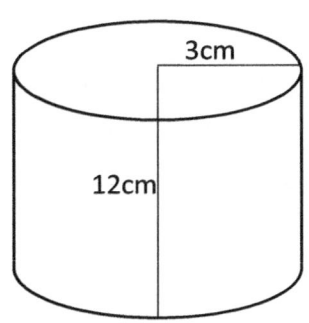
V:..............................:

2)
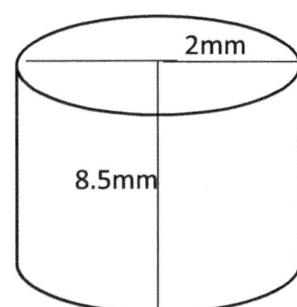
V:..............................:

3)
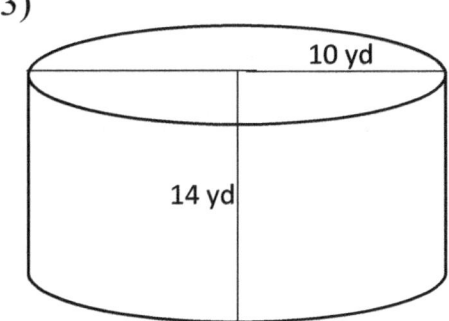
V:..............................:

4)
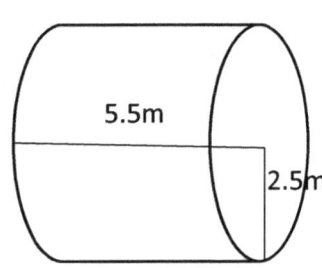
V:..............................:

5)
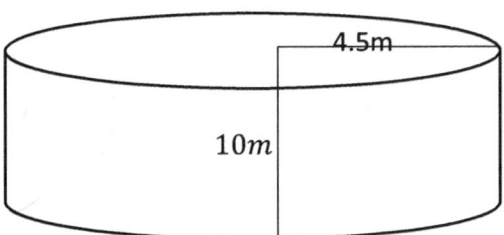
V:..............................:

6)
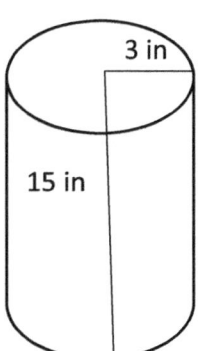
V:..............................:

Volume of Spheres

Find the volume of each spheres ($\pi = 3.14$).

1)
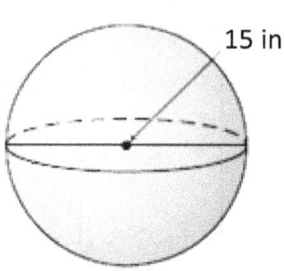

V:_____.

2)
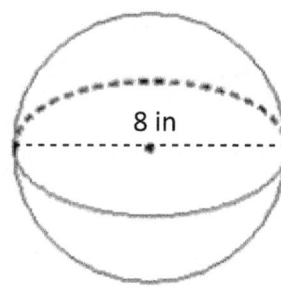

V:_____.

3)
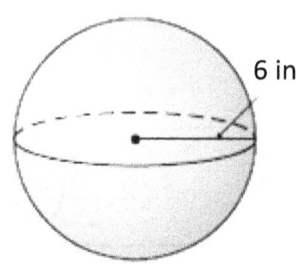

V:_____.

4)
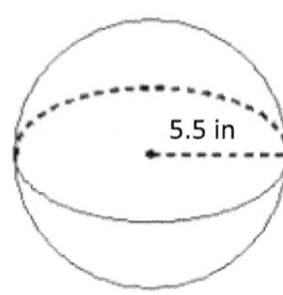

V:_____.

5)
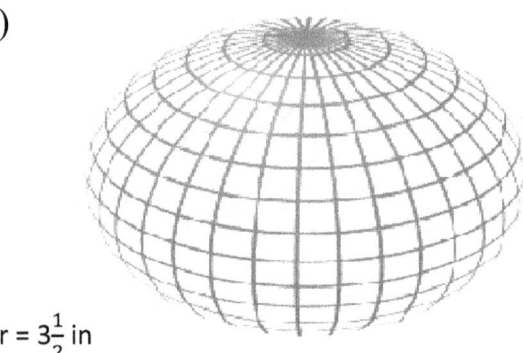

$r = 3\frac{1}{2}$ in

V:_____.

6)
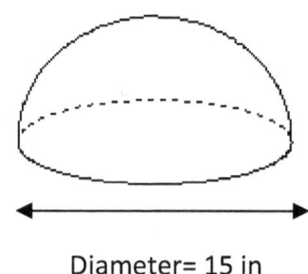

Diameter = 15 in

V:_____.

Volume of Pyramid and Cone

Find the volume of each pyramid and cone ($\pi = 3.14$).

1)

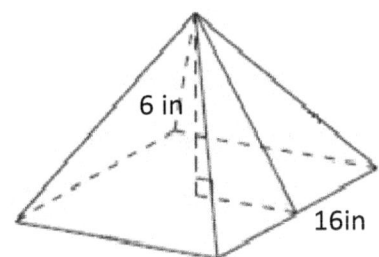

V:_____.

2)

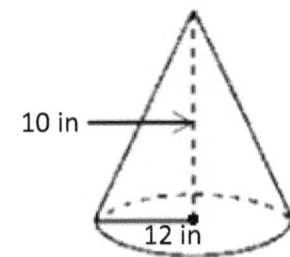

V:_____.

3)

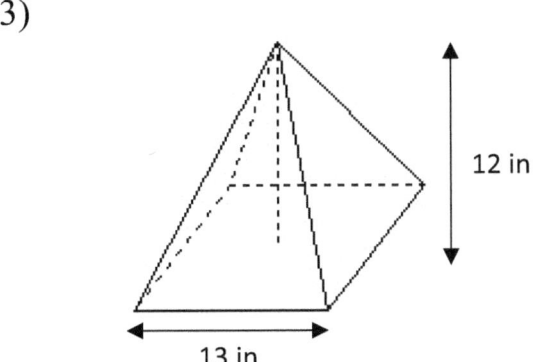

V:_____.

4)

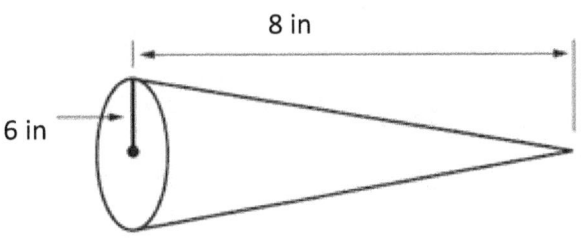

V:_____.

5)

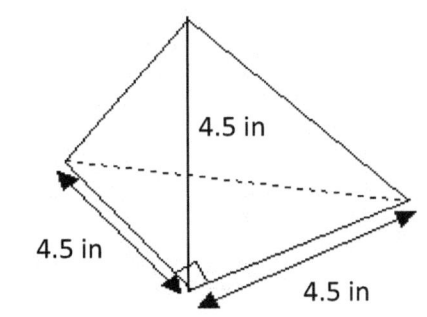

V:_____.

6)

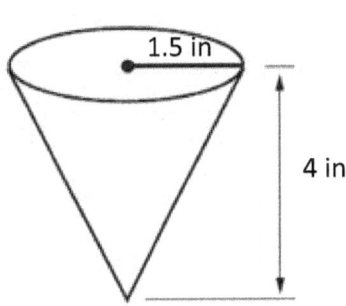

V:_____.

Surface Area Cubes

Find the surface area of each cube.

1)

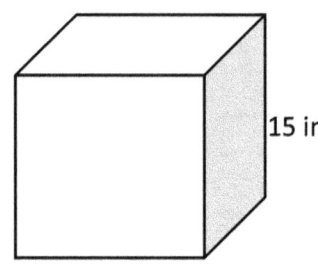

SA:_____.

2)

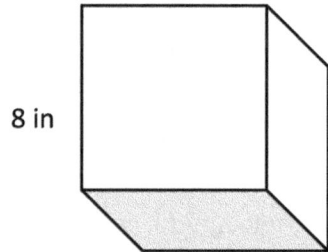

SA:_____.

3)

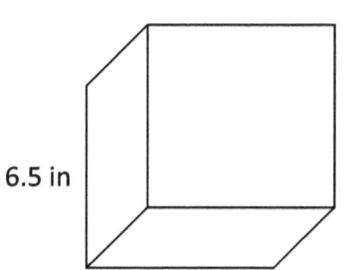

SA:_____.

4)

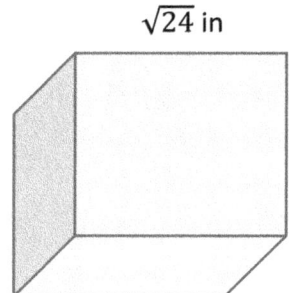

SA:_____.

5)

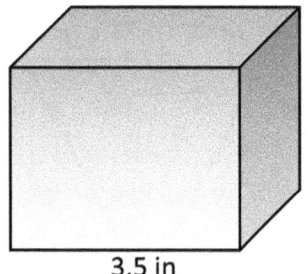

SA:_____.

6)

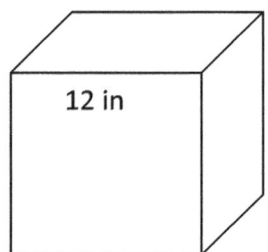

SA:_____.

Surface Area Rectangle Prism

Find the surface area of each rectangular prism.

1)
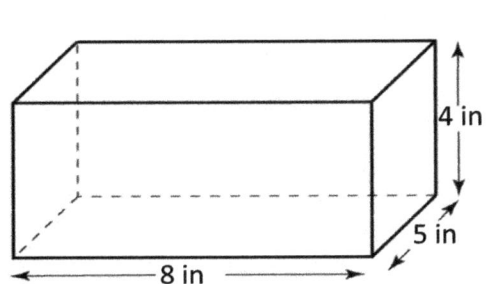

SA: _____ .

2)
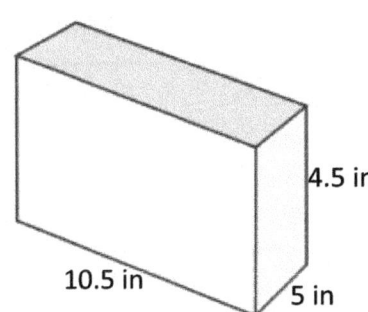

SA: _____ .

3)
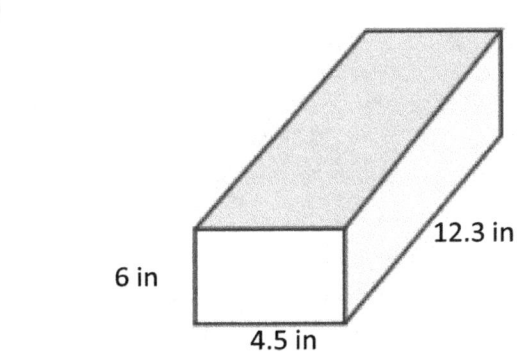

SA: _____ .

4)
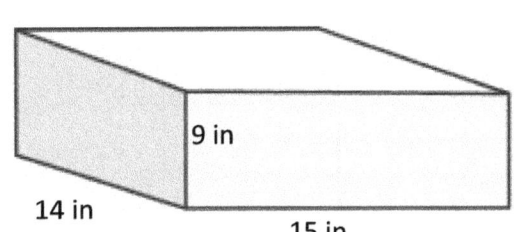

SA: _____ .

5)
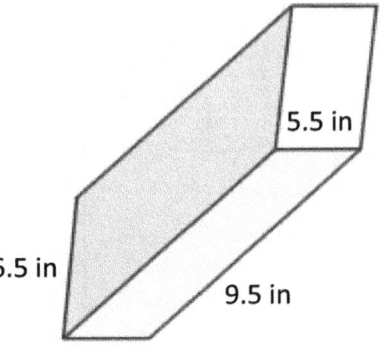

SA: _____ .

6)
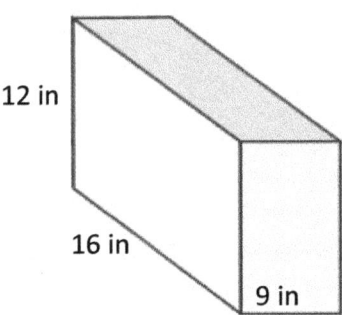

SA: _____ .

Surface Area Cylinder

Find the surface area of each cylinder.

1)
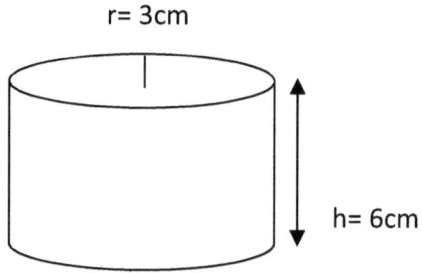

SA: ....

2)
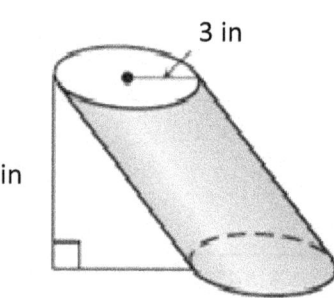

SA: ....

3)
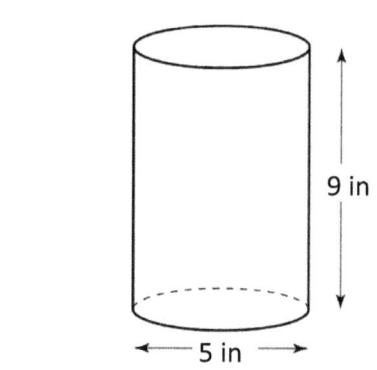

SA: ....

4)
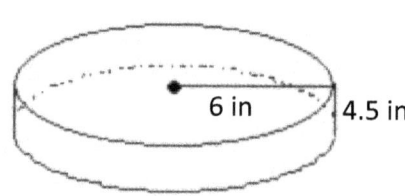

SA: ....

5)
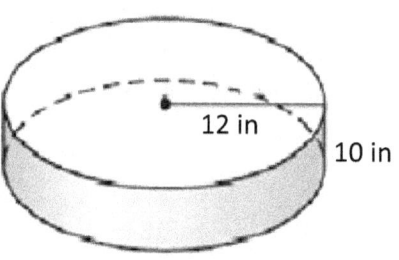

SA: ....

6)
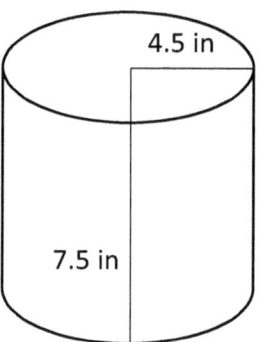

SA: ....

Answer key Chapter 10

Area and Perimeter of Square

1. Perimeter: 20,   Area:25
2. Perimeter: $4\sqrt{6}$, Area:6
3. Perimeter: 28,   Area:49
4. Perimeter: $4\sqrt{9}$, Area:9
5. Perimeter: 48,   Area:144
6. Perimeter: $4\sqrt{50}$, Area:50

Area and Perimeter of Rectangle

1- Perimeter: 18,   Area:18
2- Perimeter: 40,   Area:96
3- Perimeter: 50,   Area:150
4- Perimeter: 19, Area: 17.5
5- Perimeter: 4.23, Area: 1
6- Perimeter: 14,   Area:10

Area and Perimeter of Triangle

1- Perimeter: 3s, Area:$\frac{1}{2}sh$
2- Perimeter: 96,   Area:384
3- Perimeter: 33,   Area:45
4- Perimeter: 24,   Area:24
5- Perimeter: 24,   Area:24
6- Perimeter: 47.9, Area:97.5

Area and Perimeter of Trapezoid

1- Perimeter: 32,   Area:49
2- Perimeter: 22,   Area:26
3- Perimeter: 44,   Area:93
4- Perimeter: 50,   Area:144
5- Perimeter: 59,   Area:245
6- Perimeter: 23,   Area:17

Area and Perimeter of Parallelogram

1- Perimeter: $26m$, Area:$20(m)^2$
2- Perimeter: $50m$, Area:$78(m)^2$
3- Perimeter: $54in$, Area:$146(in)^2$
4- Perimeter: $37cm$, Area:$63(cm)^2$
5- Perimeter: $85m$, Area:$441(m)^2$
6- Perimeter: $48m$, Area:$144(m)^2$

Circumference and Area of Circle

1) Circumference:50.24 mm   Area:$200.96(mm)^2$
2) Circumference: 15.7in   Area:$(19.63in)^2$
3) Circumference: 26.38 m   Area:$55.39(m)^2$
4) Circumference: 37.68 cm   Area:113.04
5) Circumference: 21.98 in   Area:$38.47(in)^2$
6) Circumference: 9.42 km   Area:$7.07(km)^2$

Perimeter of Polygon

1) 65 mm
2) 30 m
3) 57 cm
4) 59.5 in
5) 45 m
6) 72 ft

Volume of Cubes

1) $125m^3$
2) $1,000(mm)^3$
3) $216in^3$
4) $2.74(cm)^3$

5) $32.77 (ft)^3$ 6) $729 (cm)^3$

Volume of Rectangle Prism

1) $720 (cm)^3$ 3) $39.6 (m)^3$ 5) $36 (mm)^3$
2) $75 (yd)^3$ 4) $256.5 (in)^3$ 6) $1.8 (in)^3$

Volume of Cylinder

1) $339.12 (cm)^3$ 3) $1,099 (yd)^3$ 5) $635.85 (m)^3$
2) $26.69 (mm)^3$ 4) $107.94 (m)^3$ 6) $423.9 (in)^3$

Volume of Spheres

1) $1,766.25 (in)^3$ 3) $904.32 (in)^3$ 5) $179.5 (in)^3$
2) $267.95 (in)^3$ 4) $696.56 (in)^3$ 6) $883.13 (in)^3$

Volume of Pyramid and Cone

1) $512 (in)^3$ 3) $676 (in)^3$ 5) $15.19 (in)^3$
2) $1507.2 (in)^3$ 4) $301.44 (in)^3$ 6) $9.42 (in)^3$

Surface Area Cubes

1) $1,350 (in)^2$ 3) $253.5 (in)^2$ 5) $73.5 (in)^2$
2) $384 (in)^2$ 4) $144 (in)^2$ 6) $864 (in)^2$

Surface Area Rectangle Prism

1) $184 (in)^2$ 3) $312.3 (in)^2$ 5) $299.5 (in)^2$
2) $244.5 (in)^2$ 4) $942 (in)^2$ 6) $888 (in)^2$

Surface Area Cylinder

1) $169.56 (in)^2$ 3) $180.55 (in)^2$ 5) $1,657.92 (in)^2$
2) $207.24 (in)^2$ 4) $395.64 (in)^2$ 6) $339.12 (in)^2$

Chapter 11:
Statistics and probability

Mean, Median, Mode, and Range of the Given Data

Find the mean, median, mode(s), and range of the following data.

1) 26, 69, 30, 27, 19, 54, 27

Mean: __, Median: __, Mode: __, Range: __

2) 8, 12, 12, 15, 18, 20

Mean: __, Median: __, Mode: __, Range: __

3) 51, 32, 29, 33, 39, 17, 25, 29, 12

Mean: __, Median: __, Mode: __, Range: __

4) 10, 7, 3, 9, 2, 4

Mean: __, Median: __, Mode: __, Range: __

5) 20, 16, 10, 19, 13, 18, 12, 9, 9, 7

Mean: __, Median: __, Mode: __, Range: __

6) 9, 17, 18, 9, 6, 18, 8, 12

Mean: __, Median: __, Mode: __, Range: __

7) 49, 48, 86, 96, 34, 64, 48, 14, 32, 64

Mean: __, Median: __, Mode: __, Range: __

8) 45, 45, 47, 88, 89

Mean: __, Median: __, Mode: __, Range: __

9) 18, 18, 28, 36, 64

Mean: __, Median: __, Mode: __, Range: __

10) 10, 8, 2, 2, 5, 8, 1

Mean: __, Median: __, Mode: __, Range: __

11) 5, 9, 3, 5, 1, 7

Mean: __, Median: __, Mode: __, Range: __

12) 6, 7, 11, 11, 12, 12, 12

Mean: __, Median: __, Mode: __, Range: __

13) 8, 8, 0, 16, 0, 8, 16

Mean: __, Median: __, Mode: __, Range: __

14) 12, 18, 20, 7, 11, 10, 12, 16

Mean: __, Median: __, Mode: __, Range: __

15) 6, 12, 15, 15, 20

Mean: __, Median: __, Mode: __, Range: __

16) 9, 9, 12, 10, 12, 8, 17

Mean: __, Median: __, Mode: __, Range: __

17) 20, 8, 6, 9, 18, 19, 9, 6

Mean: __, Median: __, Mode: __, Range: __

18) 62, 16, 16, 28, 3, 2

Mean: __, Median: __, Mode: __, Range: __

19) 55, 22, 24, 55, 2, 4

Mean: __, Median: __, Mode: __, Range: __

20) 98, 64, 73, 86, 91, 98, 79

Mean: __, Median: __, Mode: __, Range: __

Box and Whisker Plot

1) Draw a box and whisker plot for the data set:

   24, 21, 22, 26, 24, 22, 26, 26, 30

2) The box-and-whisker plot below represents the math test scores of 20 students.

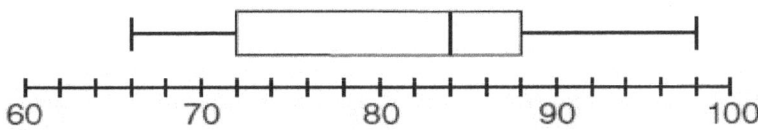

   A. What percentage of the test scores are less than 72?

   B. Which interval contains exactly 50% of the grades?

   C. What is the range of the data?

   D. What do the scores 66, 84, and 98 represent?

   E. What is the value of the lower and the upper quartile?

   F. What is the median score?

Bar Graph

Each student in class selected two games that they would like to play. Graph the given information as a bar graph and answer the questions below:

Game	Votes
Football	12
Volleyball	9
Basketball	15
Baseball	19
Tennis	15

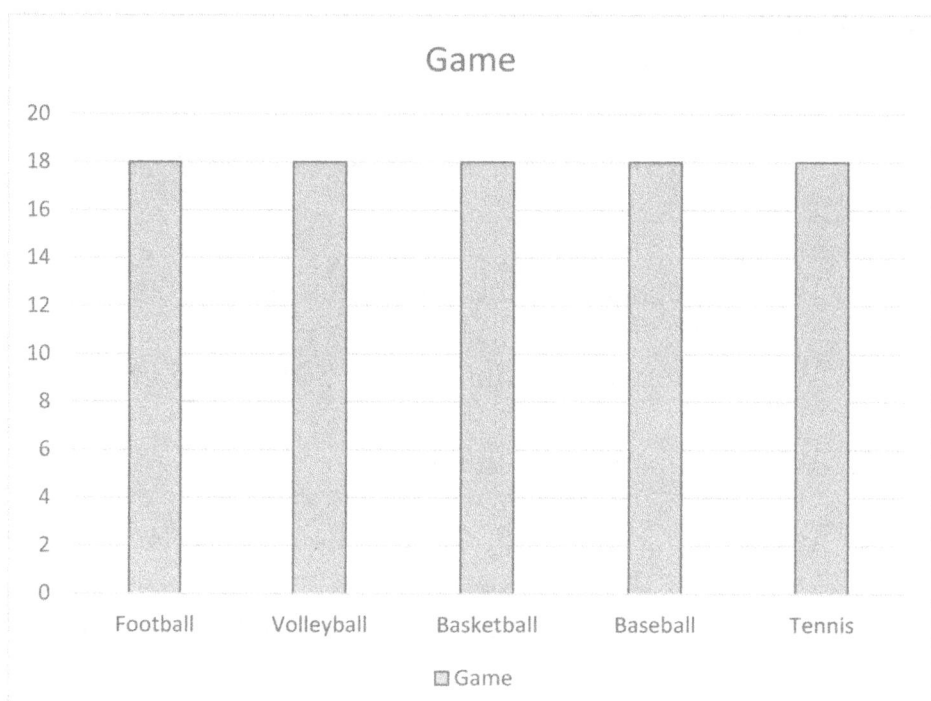

1) Which was the most popular game to play?

2) How many more student like Baseball than Football?

3) Which two game got the same number of votes?

4) How many Volleyball and Football did student vote in all?

5) Did more student like football or Tennis?

6) Which game did the fewest student like?

Dot plots

The ages of students in a Math class are given below.

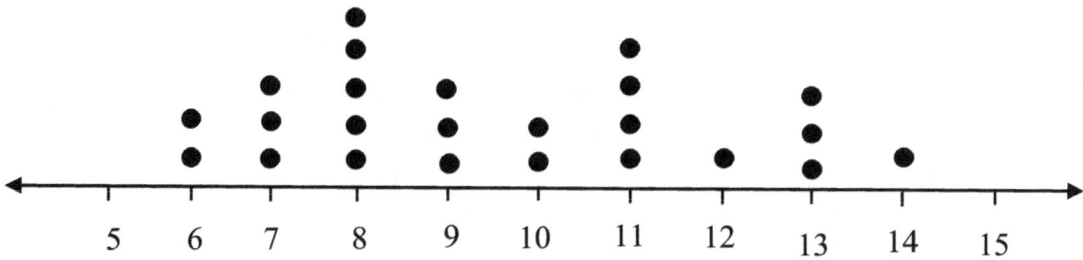

1) What is the total number of students in math class?

2) How many students are at least 11 years old?

3) Which age(s) has the most students?

4) Which age(s) has the fewest student?

5) Determine the median of the data.

6) Determine the range of the data.

7) Determine the mode of the data.

Scatter Plots

A person charges an hourly rate for his services based on the number of hours a job takes.

Hours	Rate
1	$25
2	$22.50
3	$21.50
4	$20

Hours	Rate
5	$19.50
6	$18
7	$17.50
8	$17

1) Draw a scatter plot for this data.

2) Does the data have positive or negative correlation?

3) Sketch the line that best fits the data.

4) Find the slope of the line.

5) Write the equation of the line using slope-intercept form.

6) Using your prediction equation: If a job takes 10 hours, what would be the hourly rate?

Stem–And–Leaf Plot

Make stem-and-leaf plots for the given data.

1) 22, 26, 28, 21, 42, 24, 48, 47, 29, 24, 19, 12, 45

Stem	leaf

2) 52, 54, 27, 31, 52, 24, 36, 58, 38, 34, 39, 32

Stem	leaf

3) 113, 106, 95, 95, 100, 115, 92, 114, 98, 112, 96, 107

Stem	leaf

4) 22, 15, 27, 21, 79, 24, 70, 77, 29, 24, 19, 12

Stem	leaf

5) 66, 69, 123, 67, 19, 126, 120

Stem	leaf

6) 112, 87, 96, 85, 110, 117, 92, 114, 88, 112, 98, 90

Stem	leaf

Pie Graph

80 people were survey on their favorite ice cream. The pie graph is made according to their responses. Answer following questions based on the Pie graph.

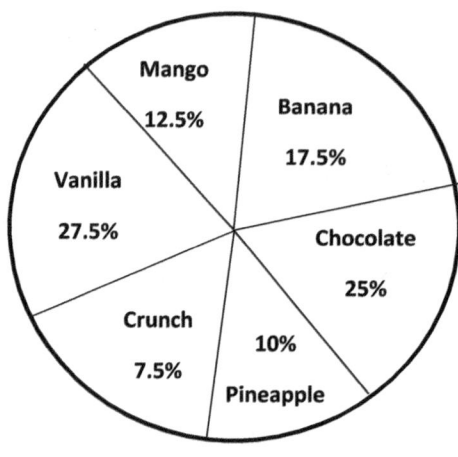

1) How many people like to eat Banana ice cream? _____

2) Approximately, which two ice creams did about half the people like the best? _____

3) How many people said either mango or crunch ice cream was their favorite? _____

4) How many people would like to have chocolate ice cream? _____

5) Which ice cream is the favorite choice of 22 people? _____

Probability

1) A jar contains 12 caramels, 7 mints and 16 dark chocolates. What is the probability of selecting a mint?

2) If you were to roll the dice one time what is the probability it will NOT land on a 2?

3) A die has sides are numbered 1 to 6. If the cube is thrown once, what is the probability of rolling a 6?

4) The sides of number cube have the numbers 3, 5, 7, 3, 5, and 7. If the cube is thrown once, what is the probability of rolling a 5?

5) Your friend asks you to think of a number from eight to twenty. What is the probability that his number will be 13?

6) A person has 5 coins in their pocket. A dime, 2 pennies, a quarter, and a nickel. If a person randomly picks one coin out of their pocket. What would the probability be that they get a penny?

7) What is the probability of drawing an odd numbered card from a standard deck of shuffled cards?

8) 24 students apply to go on a school trip. Three students are selected at random. what is the probability of selecting 3 students?

Answer key Chapter 11

Mean, Median, Mode, and Range of the Given Data

1) mean: 36, median: 27, mode: 27, range: 50
2) mean: 14.17, median: 13.5, mode: 12, range: 12
3) mean: 29.7, median: 29, mode: 29, range: 39
4) mean: 5.83, median: 5.5, mode No mode. range: 8
5) mean: 13.3, median: 12.5, mode: 9, range: 13
6) mean: 12.125, median: 10.5, mode: 9,18, range: 12
7) mean: 53.5, median: 48.5, mode: 48 and 64, range: 82
8) mean: 62.8, median: 47, mode: 45, range: 44
9) mean: 32.8, median: 28, mode: 18, range: 46
10) mean: 5.1, median: 5, mode: 2,8, range: 9
11) mean: 5, median: 5, mode: 5, range: 8
12) mean: 10.14, median: 11, mode: 12, range: 6
13) mean: 8, median: 8, mode: 8, range: 16
14) mean: 13.25, median: 12, mode: 12, range: 13
15) mean: 13.6, median: 15, mode: 15, range: 14
16) mean: 11, median: 10, mode: 9,12, range: 9
17) mean: 11.88, median: 9, mode: 6,9, range: 14
18) mean: 21.17, median: 16, mode: 16, range: 60
19) mean: 27, median: 23, mode: 55, range: 53
20) mean: 84.14, median: 86, mode: 98, range: 34

Box and Whisker Plot

1)

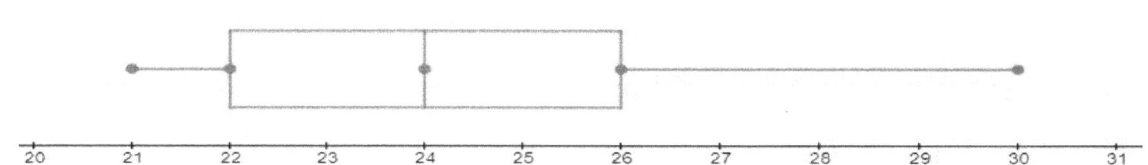

2)
- A. 25%
- B. 72-84
- C. 32
- D. Minimum, Median, and Maximum

E. Lower (Q_1) is 72 and upper (Q_3) is 88 F. 84

Bar Graph

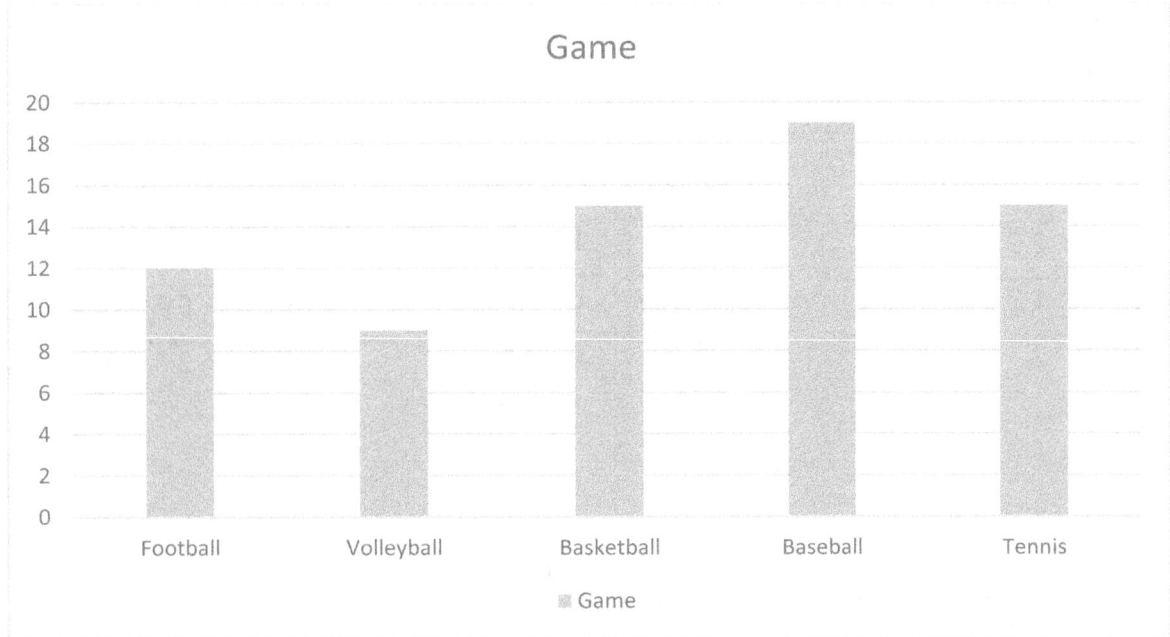

1) Baseball
2) 7 students
3) Basketball and Tennis
4) 21
5) Tennis
6) Volleyball

Dot plots

1) 24
2) 9
3) 8
4) 12 and 14
5) 3
6) 8
7) 3

Scatter Plots

1)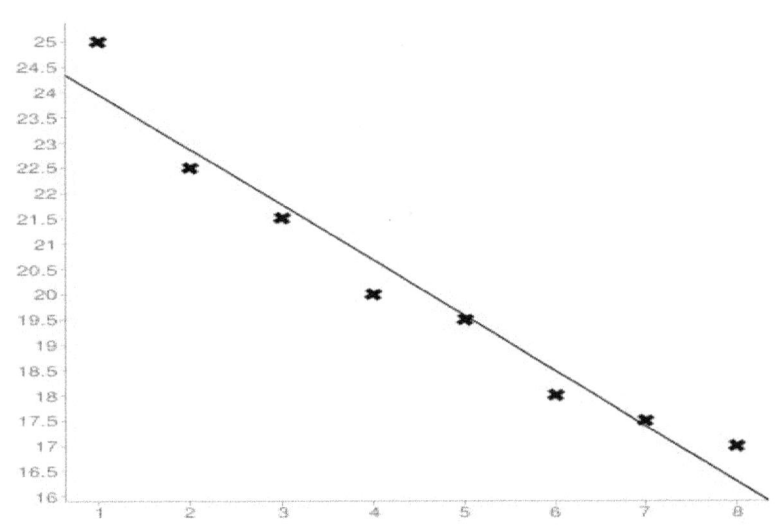

SBAC Math Practice Grade 7

2) Negative correlation

3) ----

4) Slope(m)= -1

5) $y = -x + 25$

6) 15

Stem–And–Leaf Plot

1)

Stem	leaf
1	2 9
2	1 2 4 4 6 8 9
4	2 7 8 5

2)

Stem	leaf
2	4 7
3	1 2 4 6 8 9
5	2 2 4 8

3)

Stem	leaf
9	2 5 5 6 8
10	0 6 7
11	2 3 4 5

4)

Stem	leaf
1	2 9 5
2	1 2 4 4 7 9
7	0 7 9

5)

Stem	leaf
1	9
6	6 7 9
12	0 3 6

6)

Stem	leaf
8	5 7 8
9	0 2 6 8
11	0 2 2 4 7

Pie Graph

1) 14

2) Vanilla and chocolate

3) 16

4) 20

5) Vanilla

Probability

1) $\frac{1}{5}$

2) $\frac{5}{6}$

3) $\frac{1}{6}$

4) $\frac{1}{3}$

5) $\frac{1}{12}$

6) $\frac{2}{5}$

7) $\frac{4}{13}$

8) $\frac{1}{8}$

SBAC Test Review

SBAC GRADE 7 MAHEMATICS REFRENCE MATERIALS

Linear Equations

Slope-intercept form $y = mx + b$

Constant of proportionality $k = \frac{y}{x}$

Circumference

Circle $C = 2\pi r$ or $C = \pi d$

Area

Triangle $A = \frac{1}{2}bh$

Rectangle or Parallelogram $A = bh$

Trapezoid $A = \frac{1}{2}h(b_1 + b_2)$

Circle $A = \pi r^2$

Volume

Prism or cylinder $V = Bh$

Pyramid or Cone $V = \frac{1}{3}Bh$

Additional Information

Pi $\pi = 3.14$ or $\pi = \frac{22}{7}$

Distance $d = rt$

Simple interest $I = prt$

Compound interest $I = p(1 + r)^t$

SBAC GRADE 7 MAHEMATICS REFRENCE MATERIALS

LENGTH

Customary

1 mile (mi) = 1,760 yards (yd)

1 yard (yd) = 3 feet (ft)

1 foot (ft) = 12 inches (in.)

Metric

1 kilometer (km) = 1,000 meters (m)

1 meter (m) = 100 centimeters (cm)

1 centimeter(cm) = 10 millimeters(mm)

VOLUME AND CAPACITY

Customary

1 gallon (gal) = 4 quarts (qt)

1 quart (qt) = 2 pints (pt.)

1 pint (pt.) = 2 cups (c)

1 cup (c) = 8 fluid ounces (Fl oz)

Metric

1 liter (L) = 1,000 milliliters (mL)

WEIGHT AND MASS

Customary

1 ton (T) = 2,000 pounds (lb.)

1 pound (lb.) = 16 ounces (oz)

Metric

1 kilogram (kg) = 1,000 grams (g)

1 gram (g) = 1,000 milligrams (mg)

Smarter Balanced Assessment Consortium

SBAC Practice Test 1

Mathematics

GRADE 7

Administered *Month Year*

SBAC Math Practice Grade 7

1) Peter paid for 7 sandwiches.

   - Each sandwich cost 10.22.

   - He paid for 6 bags of fries that each cost $1.87

   Which equation can be used to determine the total amount, y, Peter paid?

   A. $y = 7(10.22) + 6(1.87)x$

   B. $y = (10.22 + 1.87)x$

   C. $y = 7(10.22) + 6(1.87)$

   D. $y = 10.22x + 6(1.87)$

2) What is the decimal equivalent of the fraction $\frac{6}{11}$?

   A. 0.54

   B. $0.4\overline{54}$

   C. $0.\overline{54}$

   D. 0.545

3) The circumference of a circle is 18π centimeters. What is the area of the circle in terms of π?

   A. 18π

   B. 81π

   C. 36π

   D. 54π

SBAC Math Practice Grade 7

4) What is the volume of rectangular prism when the two triangular prisms below are stuck together?

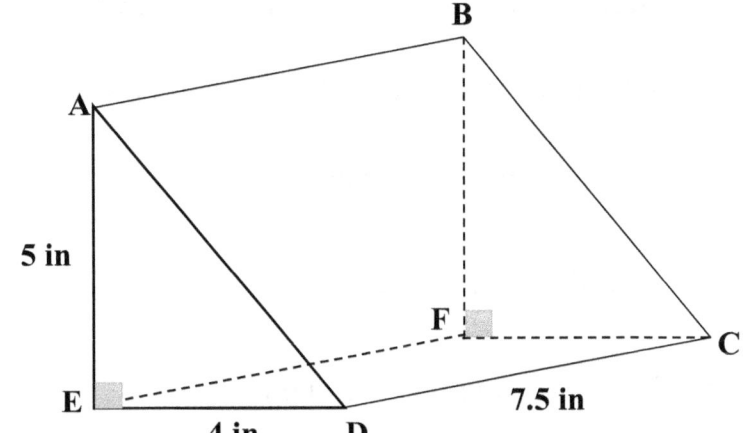

A. $150\ in^3$

B. $75\ in^3$

C. $56\ in^3$

D. $7.5\ in^3$

5) Which number line Shows the solution to the inequality $-7x - 3 < -10$?

A.

B.
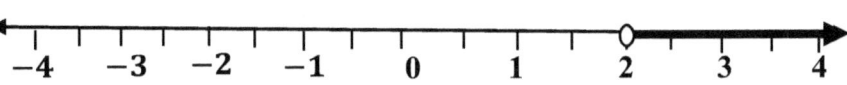

C.
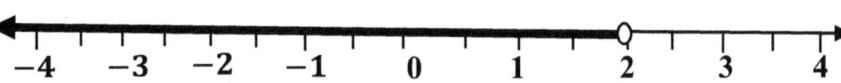

D.
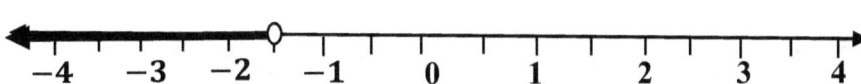

6) What is the value of $(3 + 8)^2 + (3 - 8)^2$?

A. -157

B. 25

C. 121

D. 146

WWW.MathNotion.com

SBAC Math Practice Grade 7

7) Arsan has $11 to spend on school supplies. The following table shows the price of each item in the school store. No sale tax is charged on these items. Which the combination of items can Arsan buy with his $11?

A. 4 Notebooks and 2 Pens

B. 3 Folders and 5 Erasers

C. 2 Notebooks and 4 Folders

D. 6 Erasers and 6 Pens.

Item	Price
Notebook	$2.25
Pen	$1.10
Eraser	$ 0.85
Folder	$2.15

8) If 18% of x is 72, what is 35% of x?

   A. 140

   B. 44.8

   C. 14.04

   D. 1.40

9) If all variables are positive, find the square root of $\frac{16x^9y^3}{81xy}$?

   A. $\frac{4}{9}x^6y$

   B. $\frac{9y}{4x^4}$

   C. $\frac{4}{9}x^4y$

   D. $4\frac{4}{5}x^3y^4$

WWW.MathNotion.com

10) Which is closest to the perimeter of the right triangle in the figure below?

   A. 17.2

   B. 16

   C. 17

   D. 27.2

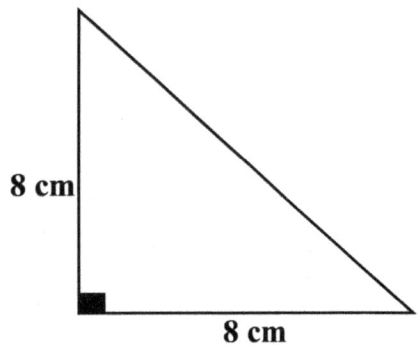

11) What is the range of the following set of data: $2, -2, 6, -1, 1, 6, 4, 3, 9$?

   A. 2

   B. 9

   C. 6

   D. 11

12) Alex starts a saving account with $60. Every week he adds $7 to his account.

   Which equation can be used to determine the number of weeks w, after which

   Alex's accounts reaches $185?

   A. $7w + 185 = 60$

   B. $7 + w = 185$

   C. $7w + 60 = 185$

   D. $7w - 60 = 185$

13) The medals won by United States, Australia and Spain during a basketball competition are shown in the table below:

United States	Australia	Spain
6	9	9

Out of the medals won by these three countries, what percentage of medals did the United States win?

A. 10%

B. 25%

C. 15%

D. 75%

14) A girl in State A spent $56 before a 6.75% sales tax and a girl in State B spent $52 before an 6.25% sales tax. How much more money did the girl from State A spend than the girl from State B after sales tax was applied? Round to the nearest hundredth.

A. 4.53

B. 45.30

C. 42

D. 15.43

15) A school has 465 students and 22 chemistry teachers and 15 physics teachers. What is the ratio between the number of physics teachers and the number of students of the school?

A. $\frac{1}{31}$

B. $\frac{9}{31}$

C. $\frac{1}{37}$

D. $\frac{15}{22}$

16) James has his own lawn mowing service. The maximum James charges to mow a lawn is $42. Which inequality represents the amount James could charge, P, to mow a lawn?

A. P < 42

B. P = 42

C. P ≤ 42

D. P ≥ 42

17) What is the value of this expression 19 ÷ 0.76?

A. 0.75

B. 4.25

C. 20

D. 25

18) The ratio of boys to girls in Maria Club is the same as the ratio of boys to girls in Hudson Club. There are 32 boys and 56 girls in Maria Club. There are 12 boys in Hudson Club. How many girls are in Hudson Club?

A. 28

B. 7

C. 18

D. 21

19) On average, Simone drinks $\frac{2}{5}$ of a 6-ounce glass of coffee in $\frac{2}{3}$ hour. How much coffee does she drink in an hour?

A. 0.7 ounces

B. 1.6 ounces

C. 3.6 ounces

D. 6.3 ounces

20) Line P, R, and S intersect each other, as shown in below diagram. Based on the angle measures, what is the value of θ?

A. 36°

B. 108°

C. 72°

D. 132°

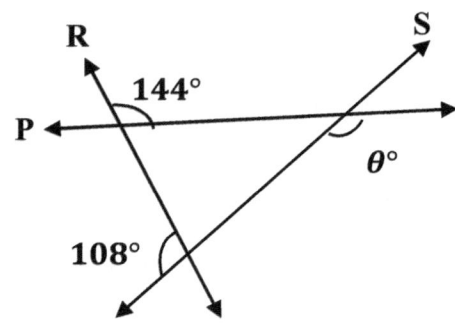

21) Which expression is represented by the model below?

A. $-4 \cdot (-9)$

B. $(-4) \cdot 9$

C. $4 \cdot (-9)$

D. $4 \cdot 9$

22) The table below shows the distance, y, a lion can travel in mile in x hour.

Time (x, hour)	Distance (y, mile)
7	364
14	728
21	1,092
28	1,456
35	1,820

Based on the information in the table, which equation can be used to model the relationship between x and y?

A. $y = x + 7$

B. $y = 7x$

C. $y = x + 364$

D. $y = 52x$

23) Mia has a loan of $42,750. This loan has a simple interest rate of 2.6% per year. What is the amount of interest that Mia will be charged on this loan at the end of one year?

A. $22,22.25

B. $11,115

C. $18,111

D. $1,111.5

24) The spinner shown has eight congruent sections.

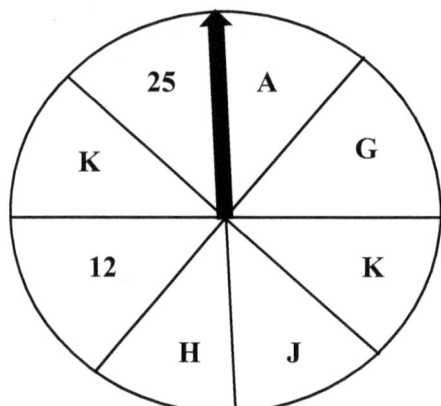

The spinner is spun 160 times. What is a reasonable prediction for the number of times the spinner will land on a letter?

A. 40

B. 12

C. 110

D. 120

25) Which graph best represents the distance a car travels when going 20 miles per hour?

A.
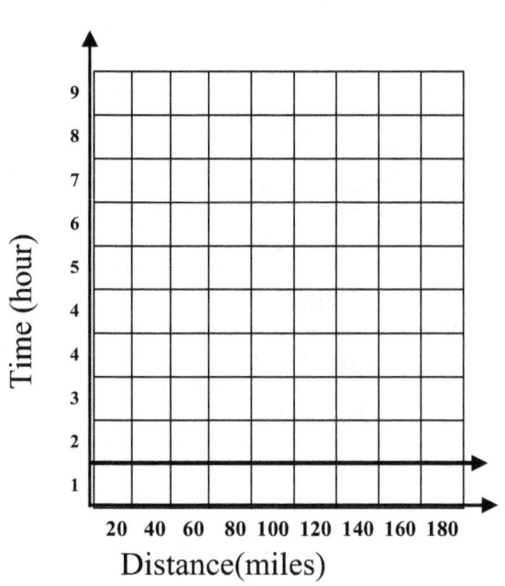

B.
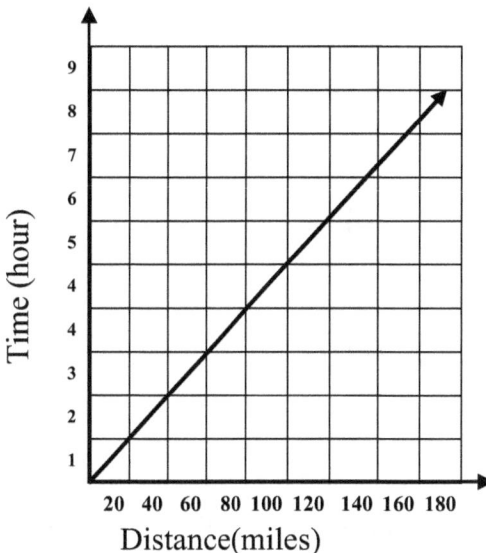

C.
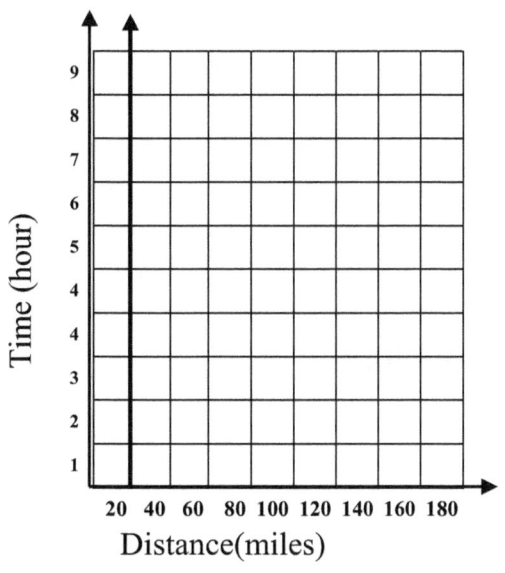

D.
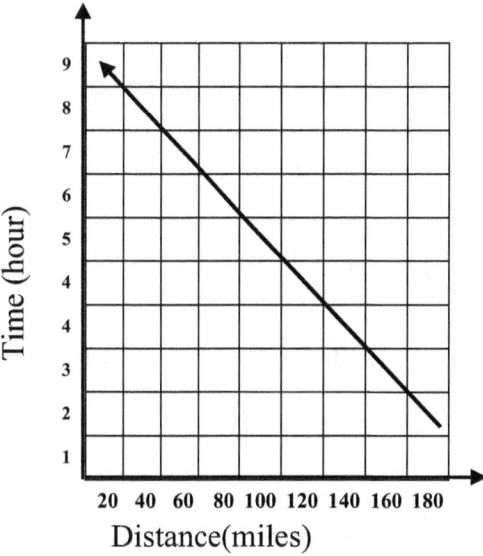

SBAC Math Practice Grade 7

26) The temperature is shown in the table below, on each of day in the week for a city in February. What is the mean temperature, in the city for that week?

A. −17

B. −19.5

C. −8

D. −7.2

Day	Temperature (°F)
Monday	−26
Tuesday	−33
Wednesday	−17
Thursday	−9
Friday	0
Saturday	11
Sunday	18

27) Which arithmetic sequence is represented by the expression $5m - 2$, where m represents the position of a term in the sequence?

A. 8, 13, 19, 24, 29, …

B. 8, 13, 18, 23, 28, …

C. 13, 18, 23, 27, 32, …

D. 13, 17, 19, 23, 28, …

28) Which expression is equivalent to $-28 - 350d$?

A. $-14(2 - 35d)$

B. $-7(40d - 7)$

C. $-14(2 + 25d)$

D. $-225\ d$

29) The dot plots show how many minutes per day do 7ᵗʰ grade study math after school at two different schools on one day.

Number of minuets study in school 1

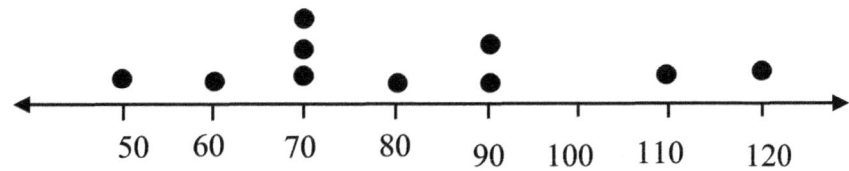

Number of minuets study in school 2

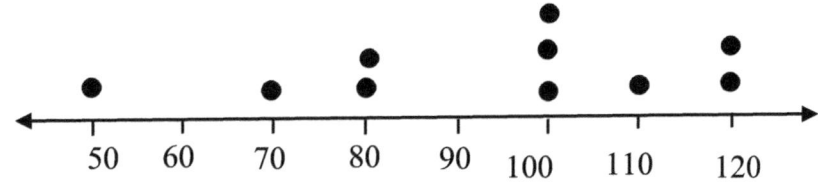

Which statement is supported by the information in the dot plots?

A. The mode of the data for School 2 is greater than the mode of the data for School 1.

B. The mean of the data for School 1 is greater than the mean of the data for School 2.

C. The median of the data for School 2 is smaller than the median of the data for School 1.

D. The median and mean of the data for two schools are equal.

30) Which number represents the probability of an event that is unlikely to occur?

A. 0.98

B. 1.5

C. 0.20

D. 0.52

SBAC Math Practice Grade 7

Smarter Balanced Assessment Consortium

SBAC Practice Test 2

Mathematics

GRADE 7

Administered *Month Year*

SBAC Math Practice Grade 7

1) What is the decimal equivalent of the fraction $\frac{47}{22}$?

   A. 2.16

   B. $2.\overline{136}$

   C. $2.1\overline{36}$

   D. 2.13

2) Thomas is shareholder of a company. The price of stock is $86.95 on the morning of day 1. Thomas records the change in the price of the stock in the chart below at the end of each day, but some information is missing.

Day	Change in Price ($)
1	+ 0.58
2	+0.76
3	
4	−0.64
5	

   The change in the price for day 3 is $\frac{3}{8}$ of the change in the price for day 4. At the end of day 5, the price of Thomas's stock is $88.29. What is the change, in dollars, in the price of the stock for day 5?

   A. −0.12

   B. 0.88

   C. −0.88

   D. 0.12

3) Kevin adds $\frac{3}{7}$ cups of sugar into a mixture every $\frac{1}{4}$ hour. What is the rate, in cups per minute, at which Kevin adds sugar to the mixture?

A. $\frac{1}{35}$

B. $2\frac{1}{7}$

C. $\frac{1}{17}$

D. $\frac{1}{105}$

4) A box of ball contains 6 blue balls, 8 red balls, 6 black balls, and 2 green balls. All the balls are the same size and shape. Brian will select a ball at random. Which of the following best describes the probability that Brian will select a green ball?

A. unlikely

B. certain

C. likely

D. impossible

5) The first number in a pattern is 5. Each following number is found by subtracting 6 from the previous number. What is the seventh number in the pattern?

A. −36

B. −21

C. −19

D. −31

SBAC Math Practice Grade 7

6) Evelyn opened a bank account. She adds the same amount of money to her account each month. The table below shows the amounts of money in her account at the ends of certain numbers of months.

   How much money does Evelyn add to her bank account each month?

   A. $9

   B. $15

   C. $18

   D. $36

Month	Amount
3	$54
6	$108
9	$162

7) Using data from house sales, probabilities for the story of a house sold were calculated. The probabilities for two story are listed below.

   - The probability a house sold has one story is 0.42.
   - The probability a house sold has two story 0.52.

   Based on these probabilities, how many of the next 500 houses sold are likely to be one story and how many are likely to be two story?

   A. one story: 42, two story: 52

   B. one story: 105, two story: 130

   C. one story: 210, two story: 260

   D. one story: 260, two story: 210

WWW.MathNotion.com

SBAC Math Practice Grade 7

8) Mr. Turner is digging a trench to put in the new school sprinkler system. Every $\frac{1}{6}$ hour, the length of his trench increases by $\frac{3}{5}$ foot. By how much does the length, in feet, of Mr. Turner's trench increase each hour?

A. $\frac{1}{5}$

B. $\frac{3}{10}$

C. $\frac{5}{18}$

D. $\frac{18}{5}$

9) Multiply: $3\frac{5}{9} \times \frac{-5}{9}$

A. $-1\frac{7}{81}$

B. $-1\frac{5}{9}$

C. $-1\frac{79}{81}$

D. -3

10) Brendan charges $26 per hour plus $70 to enter data. He accepted a project for no more than $730. Which inequality can be used to determine all the possible numbers of hours (x) it took the man to enter the data?

A. $26x + 70 \leq 730$

B. $26x + 70 > 730$

C. $70x + 26 < 730$

D. $70x + 26 \geq 730$

SBAC Math Practice Grade 7

11) Use the coordinate grid below to answer the question. What is the circumference of the circle?

A. 12.56

B. 25.12

C. 36.24

D. 169.12

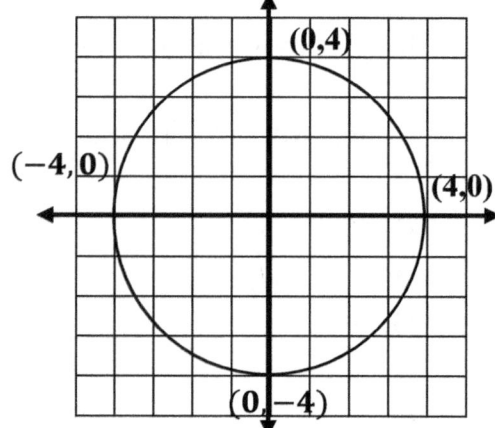

12) The temperature is 7° F. As a cold front move in, the temperature drops 5° F per half hour. What is the temperature at the end of 2 hours?

A. 13°F

B. 20°F

C. −20°F

D. −13°F

13) A printer originally cost h dollars, including tax. Eddy purchased the printer when it was on sale for 26% off its original cost. Which of the following expressions represents the final cost, in dollars, of the printer Eddy purchased?

A. $h + 0.74$

B. $h - 0.26$

C. $0.74h$

D. $0.26h$

WWW.MathNotion.com

14) Use the set of data below. What is the median of the list of numbers?

38, 23, 35, 28, 23, 30

A. 35

B. 29

C. 28

D. 30

15) Asher worked out at a gym for 4 hours. His workout consisted of jogging for 58 minutes, playing volleyball for 62 minutes, and playing billiards for the remaining amount of time. What percentage of Asher's workout was spent playing billiards?

A. 50%

B. 45%

C. 15%

D. 55%

16) The angle measures of a triangle GBD are shown in the diagram. What is the value of ∠B?

A. 19

B. 72

C. 95

D. 118

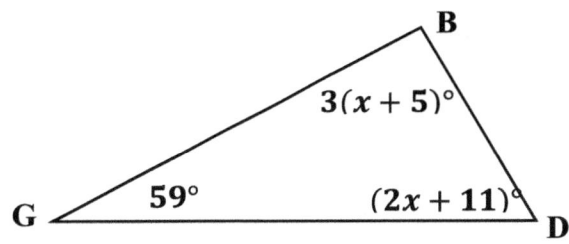

17) Triangle PRS is shown on the grid below

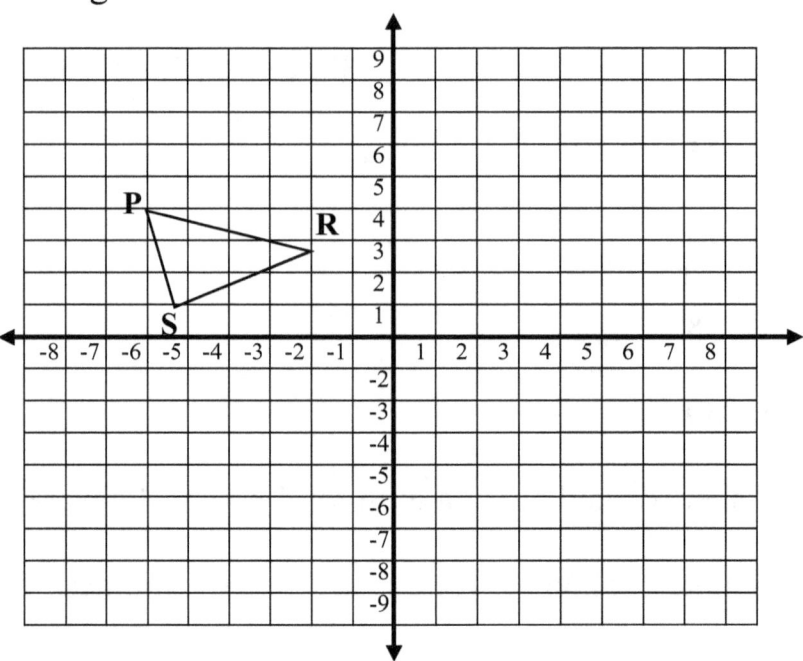

If triangle PRS is reflected across the y-axis to form triangle P'R'S', which ordered pair represents the coordinates of P'?

A. $(-6, -4)$

B. $(6, -4)$

C. $(6, 4)$

D. $(4, -6)$

18) What is the solution set for the inequality $-7x + 30 > -12$?

A. $x > 6$

B. $x < 6$

C. $x > -5$

D. $x < -5$

19) In a party people drink 118.28 liters of juices. There are approximately 29.57 milliliters in 1 fluid ounce. Which measurement is closest to the number of fluid ounces in 118.28 liters?

A. 0.004 fl oz

B. 2,782.48 fl oz

C. 2,008.84 fl oz

D. 4,000 fl oz

20) The dimensions of a square pyramid are shown in the diagram. What is the volume of the square pyramid in cubic inches?

A. 396.75 in^3

B. 396 in^3

C. 693.5 in^3

D. 961.75 in^3

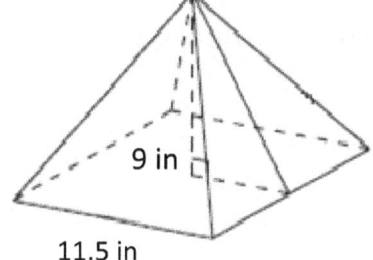

21) Water is poured to fill a pool in the shape of a rectangular prism. The pool is 15 feet long, 6.8 feet wide, and 12.7 feet high. How much cubic feet of water are needed to fill the pool?

A. 1,295.4 ft^3

B. 129.45 ft^3

C. 129.54 ft^3

D. 129.94 ft^3

SBAC Math Practice Grade 7

22) The store manager spent $13,720 to buy a new freezer and 28 tables. The total purchase is represented by this equation, where v stands for the value of each table purchased: $28v + 1,120 = 13,720$

What was the cost of each table that the manager purchased?

A. $505

B. $500

C. $544

D. $450

23) In a city, at 2:15 A.M., the temperature was $-6°F$. At 2:15 P.M., the temperature was $15°F$. Which expression represents the increase in temperature?

A. $-6 - 15$

B. $|-6 - 15|$

C. $|-6| - 15$

D. $-6 - |15|$

24) Angles α and β are complementary angles. Angles α and are supplementary angles. The degree measure of angle β is $110°$. What is the measure of angle γ?

A. 20°

B. 110°

C. 70°

D. 60°

25) The bar graph shows a company's income and expenses over the last 5 years.

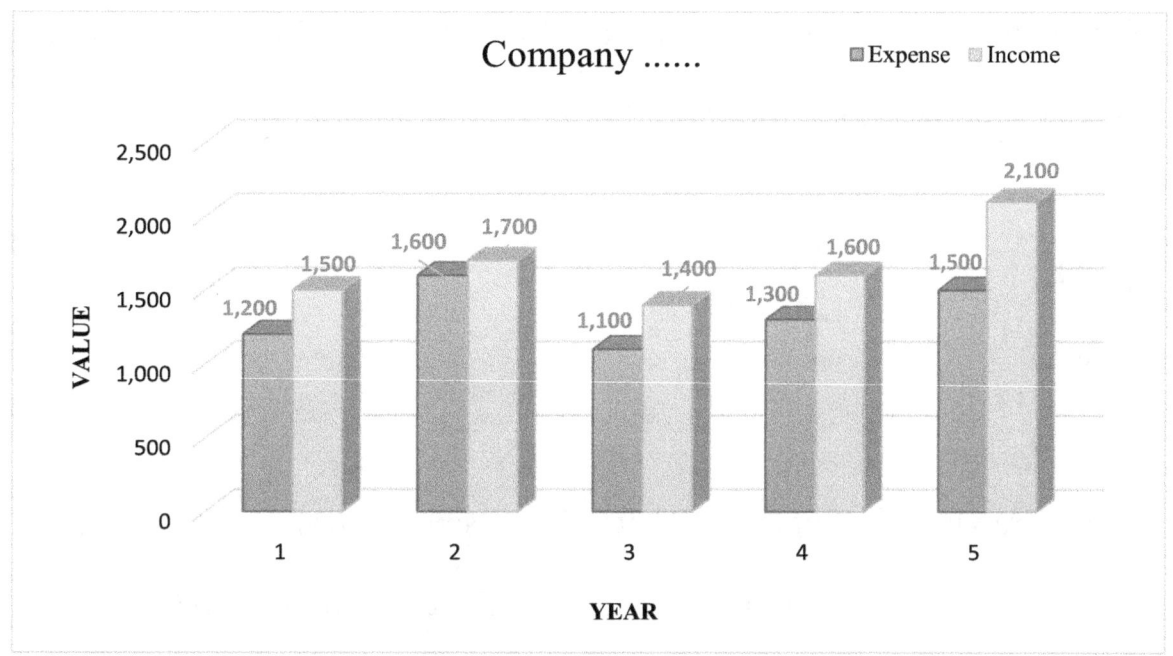

Which statement is supported by the information in the graph?

A. Expenses have increased $400 each year over the last 5 years.

B. The income in Year 5 was 15% more than the income in Year 1.

C. The combined income in Years 3, 4, and 5 was equal to the combined expenses in Years 2, 3, and 4.

D. Expenses in the year 3 was more than half of the income in the year 4.

26) Which expression is equivalent to the $(3n - 9) - \frac{1}{3}(8 - 9n) + \frac{5}{3}$?

A. -6

B. $-3n - 10$

C. $6n - 10$

D. $6n - 6$

27) Patricia bought a bottle of 16-ounce balsamic vinegar for $13.06. She used 35% of the balsamic vinegar in two weeks. Which of the following is closest to the cost of the balsamic she used?

A. $0.45

B. $7.47

C. $4.57

D. $6.75

28) A scale drawing of triangle DEF that will be used on a wall is shown below. What is the perimeter, in meter, of the actual triangle used on the wall?

Scale: 1 cm: $3\frac{1}{5}$ m

A. $21\frac{2}{5}$

B. $57\frac{3}{5}$

C. $67\frac{1}{5}$

D. $47\frac{1}{5}$

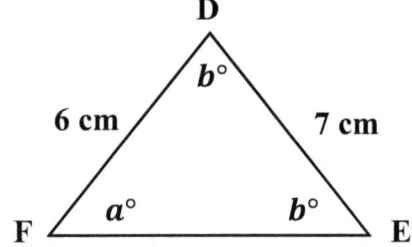

29) The ratio of boys to girls in Geometry class is 4 to 5. There are 25 girls in the class. What is the total number of students in Geometry class?

A. 65

B. 45

C. 20

D. 15

30) A group of employees have their weight recorded to make a data set. The mean, median, mode, and range of the data set are recorded. Then, the weight of the manager is included to make a new data set. The manager's weight is more than all but one of the employees. Which measure must be the same when the manager's weight included?

A. Mean

B. Mode

C. Median

D. Range

Answers and Explanations

Answer Key

Now, it's time to review your results to see where you went wrong and what areas you need to improve!

SBAC Math Practice Tests

Practice Test 1

#	Ans	#	Ans	#	Ans
1	C	11	D	21	C
2	C	12	C	22	D
3	B	13	B	23	D
4	A	14	A	24	D
5	A	15	A	25	B
6	D	16	C	26	C
7	C	17	D	27	B
8	A	18	D	28	C
9	C	19	C	29	A
10	D	20	B	30	C

Practice Test 2

#	Ans	#	Ans	#	Ans
1	C	11	B	21	A
2	B	12	D	22	D
3	A	13	C	23	B
4	A	14	B	24	A
5	D	15	A	25	D
6	C	16	B	26	C
7	C	17	C	27	C
8	D	18	B	28	C
9	C	19	D	29	B
10	A	20	A	30	D

SBAC Practice Test 1
Answers and Explanations

1) Answer: C

Let y be the total amount paid.

We have been given that Peter bought 7 sandwiches that each cost the $10.22.

So, the cost of 7 sandwiches would be 7(10.22). He paid for 6 bags of fries that each cost the $1.87. So, the cost of 6 bags would be 6(1.87)

Then, the total cost of sandwiches and fries would be $y = 7(10.22) + 6(1.87)$

2) Answer: C

To get the answer to 6 over 11 as a decimal, we divide 6 by 11.

$\frac{6}{11} = 0.545454 \ldots = 0.\overline{54}$

3) Answer: B

Use the formula of circumference of circles.

Circumference = $\pi d = 2\pi (r) = 18 \pi \Rightarrow r = 9$

Radius of the circle is 9. Now, use the areas formula:

Area = $\pi r^2 \Rightarrow$ Area = $\pi(9)^2 \Rightarrow$ Area = 81π

4) Answer: A

The volume of a triangular prism is the base times the height. $V = Bh$

Area of the base = $\frac{1}{2} b.h \rightarrow B = \frac{1}{2} \times 4 \times 5 = 10$

$V = B.h = 10 \times 7.5 = 75$ and we need two triangular prisms, then $2 \times 75 = 150$

5) Answer: A

$-7x - 3 < -10$, add 3 to both sides $-7x < -7$ divide each term by -7

If an inequality is multiplied or divided by a negative number, you must change the direction of the inequality, then $x > 1$

SBAC Math Practice Grade 7

6) Answer: D

$(3+8)^2 + (3-8)^2 = (11)^2 + (-5)^2 = 121 + 25 = 146$

7) Answer: B

A. $(4 \times 2.25) + (2 \times 1.10) = 9 + 2.20 = 11.2 > 11$

B. $(3 \times 2.15) + (5 \times 0.85) = 6.45 + 4.25 = 10.7 < 11$

C. $(2 \times 2.25) + (4 \times 2.15) = 4.5 + 8.6 = 13.1 > 11$

D. $(6 \times 0.85) + (6 \times 1.10) = 5.1 + 6.6 = 11.7 > 11$

8) Answer: A

$0.18 \times x = 72 \rightarrow x = \frac{72}{0.18} = \frac{7,200}{18} = 400$

$35\% \, of \, 400 = 0.35 \times 400 = 140$

9) Answer: C

$\sqrt{\frac{16x^9y^3}{81xy}} = \sqrt{\frac{16}{81} \times \frac{x^9y^3}{xy}} = \sqrt{\frac{16}{81}x^8y^2} = \frac{4}{9}x^4y$

10) Answer: D

Use Pythagorean Theorem: $a^2 + b^2 = c^2$

$8^2 + 8^2 = c^2 \Rightarrow 64 + 64 = c^2 \Rightarrow 128 = c^2 \Rightarrow c = \sqrt{128} = 8\sqrt{2} = 8 \times 1.4 = 11.2$

Perimeter of triangle $= a + b + c = 8 + 8 + 11.2 = 27.2$

11) Answer: D

Range= Largest Value – Smallest Value $\rightarrow R = 9 - (-2) = 11$

12) Answer: C

To start, assign the variables to unknowns, known values to constants, and relate them by the relations between the variables and constants. Then

$7w + 60 = 185$

13) Answer: B

Use percent formula: part $= \frac{percent}{100} \times$ whole

SBAC Math Practice Grade 7

Whole $= 6 + 9 + 9 = 24$

$6 = \frac{\text{percent}}{100} \times 24 \Rightarrow 6 = \frac{\text{percent} \times 24}{100} \Rightarrow 600 = \text{percent} \times 24 \Rightarrow \text{percent} = \frac{600}{24} = 25$,

Therefore United States win 25% of medals.

14) Answer: A

Multiply the price by the sales tax to find out how much money the sales tax will add, then Add the original price and the sales tax.

State A: $56 \times 0.0675 = 3.78$

$56 + 3.78 = 59.78$

State B: $52 \times 0.0625 = 3.25$

$52 + 3.25 = 55.25$

Then take the difference: $59.78 - 55.25 = 4.53$

15) Answer: A

15 physics teachers to 465 students is 15:465, 1:31

16) Answer: C

At least and Minimum – means greater than or equal to

At most, no more than, and Maximum – means less than or equal to

More than – means greater than

Less than – means less than

Then, $P \leq 42$

17) Answer: D

$\frac{19}{0.76}$ Multiply the numerator and denominator by 100: $\frac{1,900}{76} = 25$

18) Answer: D

The ratio of boys to girls in Maria Club: $32:56 = 4:7$

The ratio of boys to girls in Hudson Club: $4:7$

4 :7 same as 12: 21.

So, there are 21 girls in Hudson club

19) Answer: C

$\frac{2}{5} \times 6 = 2.4$

$\frac{2}{3} \times 60 = 40 \; min$

$\frac{2.4}{40} = \frac{x}{60} \rightarrow 40x = 2.4 \times 60 \rightarrow x = \frac{144}{40} = 3.6$ Ounces

20) Answer: B

Supplementary angles are two angles that have a sum of 180°

$ine \; R: 180° - 144° = 36°, Line \; S: 180° - 108° = 72°$

then in the triangle: $180° - (72° + 36°) = 72°$

$line \; P: \theta° = 180° - 72° = 108°$

21) Answer: C

$4 \times (-9) = -36$

22) Answer: D

$\frac{364}{7} = 52$

$\frac{728-364}{14-7} = \frac{364}{7} = 52$

23) Answer: D

$42,750 \times 0.026 = 1,111.5$

24) Answer: D

The diagram shows a spinner with 8 sections. The probability land on letter is $\frac{6}{8}$

The spinner is spun 160 times, then the prediction is: $160 \times \frac{6}{8} = 120$ times

25) Answer: B

A linear equation is a relationship between two variables, and application of linear equations can be found in distance problems.

$d = rt$ or distance equals rate (speed) times time.

$d = 1 \times 20 = 20$, then $(1,20), (2,40), (3,60)(4,80), \ldots$

SBAC Math Practice Grade 7

26) Answer: C

average (mean) $=\frac{sum\ of\ terms}{number\ of\ terms}=\frac{(-26)+(-33)+(-17)+(-9)+0+11+18}{7}=\frac{-56}{7}=-8°F$

27) Answer: B

$m = 1 \rightarrow 5m - 2 = 5(1) - 2 = 3$

$m = 2 \rightarrow 5m - 2 = 5(2) - 2 = 8$

$m = 3 \rightarrow 5m - 2 = 5(3) - 2 = 13$

$m = 4 \rightarrow 5m - 2 = 5(4) - 2 = 18$

$m = 5 \rightarrow 5m - 2 = 5(5) - 2 = 23$

28) Answer: C

$-28 - 350d = -14(2 + 25d)$

29) Answer: A

Let's find the mode, mean (average), and median of the number of minutes for each school.

Number of Minutes for school 1: 50, 60, 70, 70, 70, 80, 90, 90, 110, 120

Mean(average) $=\frac{sum\ of\ terms}{number\ of\ terms}=\frac{50+60+70+70+70+80+90+90+110+120}{10}=\frac{810}{10}=81$

Median is the number in the middle. Since there are an even number of items in the resulting list, the median is the average of the two middle numbers.

Median of the data is $(70 + 80) \div 2 = 75$

Mode is the number which appears most often in a set of numbers. Therefore, there is no mode in the set of numbers. Mode is: 70

Number of Minutes for school 2: 50, 70, 80, 80, 100, 100, 100, 110, 120, 120

Mean $=\frac{50+70+80+80+100+100+100+110+120+120}{10}=\frac{930}{10}=93$

Median: $(100 + 100) \div 2 = 100$

Mode: 100

30) Answer: C

We often describe the probability of something happening with words like impossible, unlikely, as likely as unlikely, equally likely, likely, and certain. The probability of an event occurring is represented by a ratio. A ratio is a number that is between 0 and 1 and can include 0 and 1. An event is impossible if it has a probability of 0. An event is certain if it has the probability of 1

impossible	unlikely	equally likely, equally unlikely	likely	Certain
0		$\frac{1}{2}$		1

SBAC Practice Test 2
Answers and Explanations

1) Answer: C

Divided 47 by 22: $\frac{47}{22} = 2.1363636\ldots = 2.1\overline{36}$

2) Answer: B

Day 3: $-0.64 \times \frac{3}{8} = -0.24$

Change price in days: $(86.95 + 0.58 + 0.76 + (-0.24) + (-0.64)) = 87.41$

Day 5: $88.29 - 87.41 = 0.88$

3) Answer: A

$1\,hour = 60\,min \rightarrow \frac{1}{4} \times 60 = 15\,min$

Rate: $\frac{\frac{3}{7}}{15} = \frac{x}{1} \rightarrow 15x = \frac{3}{7} \rightarrow x = \frac{3}{105} = \frac{1}{35}$

4) Answer: A

Probability $= \frac{number\ of\ desired\ outcomes}{number\ of\ total\ outcomes} = \frac{2}{6+8+6+2} = \frac{2}{22} = \frac{1}{11}$

If an event has a 0 probability this means that it can never happen

If an event has a 1 probability it will certainly happen

If an event has a 0.5 probability, it has an equal chance of happening or not happening

If an event has a probability between 0 and 0.5, then it is unlikely to happen

If an event has a probability between 0.5 and 1, then it is likely to happen,

5) Answer: D

The pattern is: $5, -1, -7, -13, -19, -25, 31$

6) Answer: C

$54 \div 3 = 18$

$108 \div 6 = 18$

SBAC Math Practice Grade 7

7) Answer: C

likely one-story house is: $0.42 \times 500 = 210$

likely of two-story house is: $0.52 \times 500 = 260$

8) Answer: D

Write the ratio and solve for x.

$\frac{1}{\frac{1}{6}} = \frac{x}{\frac{3}{5}}$ (Cross multiply) $\Rightarrow \frac{1}{6}x = \frac{3}{5} \Rightarrow x = \frac{18}{5}$

9) Answer: C

$3\frac{5}{9} \times \frac{-5}{9} = \frac{32}{9} \times \frac{-5}{9} = -\frac{160}{81} = -1\frac{79}{81}$

10) Answer: A

Hour: $x \rightarrow 26$ per hour: $26x$

Plus: add $(+)$, no more than: $\leq$; Then, $26x + 70 \leq 730$

11) Answer: B

By grid line: $d = 8$

Or distance for two points $(0,4), (0,-4)$: $d = \sqrt{(0-0)^2 + (4-(-4))^2} = 8$

$d = 8 \rightarrow$ Circumference $= \pi d = \pi(8) = 8\pi = 25.12$

12) Answer: D

2 hours equal 4 half hours

$4 \times 5° = 20° \rightarrow 7 - 20 = -13°F$

13) Answer: C

If the price of a printer is decreased by 38% then: $100\% - 26\% = 74\%$

74% of $h = 0.74 \times h = 0.74h$

14) Answer: B

The median of a set of data is the value located in the middle of the data set. To find median, first list numbers in order from smallest to largest:

23, 23, 28, 30, 35, 38

Since there are an even number of items in the resulting list, the median is the average of the two middle numbers.

Median= $(28 + 30) \div 2 = 29$

15) Answer: A

Each hour is 60 minutes, so we have $4 \times 60 = 240$ minutes of workout time for Asher. We subtract off the jogging and playing volleyball time to get the time Asher playing billiards: $240 - 58 - 62 = 120$ minutes

$percent = \left(\frac{part}{whole}\right) \times 100 \rightarrow percent = \left(\frac{120}{240}\right) \times 100 \rightarrow percent = 50\%$

16) Answer: B

$3(x + 5) + (2x + 11) + 59 = 180 \rightarrow 3x + 15 + 2x + 11 + 59 = 180$

$5x + 85 = 180 \rightarrow 5x = 95 \rightarrow x = \frac{95}{5} = 19 \Rightarrow x = 19°$

$\angle B = 3(x + 5°) = 3(19° + 5°) = 72°$

17) Answer: C

The reflection of the point (x, y) across the y-axis is the point $(-x, y)$.

If you reflect a point across the y-axis, the y-coordinate is the same, but the x-coordinate is changed into its opposite.

Reflection of $(-6, 4) \rightarrow (6, 4)$

18) Answer: B

$-7x + 30 > -12$ (Subtract 30 from both sides)

$-7x > -42$ (Divide both side by -7, remember negative change the sign)

$x < 6$

19) Answer: D

1 L=1,000 mL

1 fl oz = 29.57 mL→ 1 fl oz =0.02957 L

$118.28 \text{ L} = \frac{118.28}{0.02957} \times \frac{100,000}{100,000} = \frac{11,828,000}{2,957} = 4,000 \: fl \: oz$

SBAC Math Practice Grade 7

20) Answer: A

Volume of a square-based pyramid: $V = \frac{1}{3} B.h$, where V is the volume and B is the area of the base. Then, $V = \frac{1}{3} \times 11.5 \times 11.5 \times 9 = \frac{1,190.25}{3} = 396.75$

21) Answer: A

Use formula of rectangle prism volume.

$V = $ (length) (width) (height) $\Rightarrow V = 15 \times 6.8 \times 12.7 = 1,295.4 \; ft^3$

22) Answer: D

$28v + 1,120 = 13,720$ (subtract 1,120 from both sides)

$\rightarrow 28v = 12,600$ (divide both sides by 28) $\rightarrow v = \$450$ cost of each table

23) Answer: B

Difference of temperature is: $|t_2 - t_1| = |15 - (-6)| = |15 + 6| = |-15 - 6|$

24) Answer: A

Supplementary angles are two angles with a sum of 180 degrees.

$\alpha + \beta = 180°$ and $\beta = 90° \Rightarrow \alpha = 180° - 110° = 70°$

complementary angles are two angles with a sum of 90 degrees.

$\alpha + \gamma = 90$ and $\alpha = 70° \Rightarrow \gamma = 90° - 70° = 20°$

25) Answer: D

A. Expenses in Year 2: $1,600 \rightarrow 1,600 + 400 = 2,000 \neq$ Year 3

B. Income in Year 1: 1,500 and 15% of $1,500 = 225 \rightarrow 1,500 + 225 = 1,725 \neq$ income of Year 5

C. Incomes year 3, 4, and 5: $1,400 + 1,600 + 2,100 = 5,100$

Expenses year 2, 4, and 5: $1,600 + 1,300 + 1,500 = 4,400 \neq 5,100$

D. Half of Year incomes 4: $\frac{1,600}{2} = 800 < 1,100$ Expenses in Year 3

26) Answer: C

$(3n - 9) - \frac{1}{3}(8 - 9n) + \frac{5}{3} = 3n - 9 - \frac{8}{3} + 3n + \frac{5}{3} = 6n - 10$

27) Answer: C

If you ever need to find the percentage of something you just times it by the fraction. So, all you need to do to figure this out is to find 35% of 13.06 which is approximately 4.57.

28) Answer: C

$\angle D = \angle E = b° \to FD = FE = 6cm$

Perimeter of triangle: $6 + 6 + 7 = 21\ cm$

Actual triangle: $21 \times 3\frac{1}{5} = 21 \times \frac{16}{5} = \frac{336}{5} = 67\frac{1}{5}\ cm$

29) Answer: B

The ratio of boy to girls is 4:5. Therefore, there are 5 girls out of 9 students. To find the answer, first divide the number of girls by 5, then multiply the result by 9.

$25 \div 5 = 5 \Rightarrow 5 \times 9 = 45$

30) Answer: D

A. does not consider that the manager's weight could be different from the mean weight and would change the value of the mean

B. does not consider that the manager could be the same weight as the employee and that this weight could be the new mode

C. does not consider that the nth heaviest employee (in the middle) could be different weight and adding a weight that is greater than either of these weights to the data set would change the value of the median

D. corrects.

"End"